DRUM GROOVE CREATION

Drum Groove Creation: For Creative Drummers and Producers provides a comprehensive guide to creating new drum grooves by using unique methods and tools.

The book covers popular genres, from Jazz and Blues to modern Hip-Hop subgenres, as well as different cultural rhythms from around the world. It is designed to equip the modern drummer, producer and beatmaker with the tools to be able to create their own drum grooves. *Drum Groove Creation* will take you from the absolute essentials of groove construction to the tools you need based on defining characteristics and musical influences. Each groove is showcased in drum set notation and in a piano roll grid system as it would appear on a DAW to ensure it can be used by drummers and producers.

This book is for anyone wanting to learn about the different grooves and musical examples they may encounter in the modern music industry, as well as producers, composers and songwriters who need to write, programme and create drums in their own work.

Hailed as "**A Revelation**" by *Prog Magazine*, **Guy Monk** is a British drummer, educator, and author from Worcestershire, England. After studying at London's renowned DrumTech, he built an international career performing across the UK, Europe, the United States, and the Caribbean. Guy holds a Master's degree in Learning and Teaching and is the founder of **The British Drumming Academy**, where he mentors drummers worldwide. Alongside touring, recording, and university lecturing, he remains dedicated to inspiring the next generation of musicians through performance and education.

I0894159

DRUM GROOVE CREATION

For Creative Drummers and Producers

Guy Monk
Images and artwork by
Jessica Monk

Routledge
Taylor & Francis Group
LONDON AND NEW YORK

Designed cover image: Guy Monk

First published 2027
by Routledge
4 Park Square, Milton Park, Abingdon, Oxon OX14 4RN

and by Routledge
605 Third Avenue, New York, NY 10158

Routledge is an imprint of the Taylor & Francis Group, an informa business

British Library Cataloguing-in-Publication Data
A catalogue record for this book is available from the British Library

Library of Congress Cataloging-in-Publication Data
Names: Monk, Guy author | Monk, Jessica illustrator artist Title:
Drum groove creation : for creative drummers & producers / Guy Monk;
images and artwork by Jessica Monk.
Description: Abingdon, Oxon ; New York : Routledge, 2026. |
Includes index.
Identifiers: LCCN 2026015630 (print) | LCCN 2026015631 (ebook) |
ISBN 9781032954936 hardback | ISBN 9781032954851 paperback |
ISBN 9781003585169 ebook
Subjects: LCSH: Drum set | Drum machine | LCGFT: Studies (Music)
Classification: LCC MT662.3 .M66 2026 (print) | LCC MT662.3 (ebook) |
DDC 786.9/193--dc23/eng/20260617
LC record available at https://lccn.loc.gov/2026015630
LC ebook record available at https://lccn.loc.gov/2026015631

ISBN: 978-1-032-95493-6 (hbk)
ISBN: 978-1-032-95485-1 (pbk)
ISBN: 978-1-003-58516-9 (ebk)

DOI: 10.4324/9781003585169

Typeset in Zurich
by SPi Technologies India Pvt Ltd (Straive)

Disclaimer

Please note that this book contains my interpretations of notable and key drum grooves from throughout the 20th and 21st century. I do not own the rights to the master recordings or claim ownership over the intellectual property discussed within the text. This book is for educational purposes.

Contents

Contents

Contents

About the author

Guy Monk, hailed as *"a revelation"* by the international publication *Prog Magazine*, rises from a proud lineage of British drummers as a drummer, educator and author. Born and raised in Worcestershire, England, his career is driven by a deep commitment to both artistic excellence and the mentorship of drummers at every stage of their journey.

Relocating to London at just 18 to study at the respected 'DrumTech', Guy turned professional at age 21, marking his arrival with the release of the debut album *The Back Room* alongside Oli Barton on Coke & Dagger Records. This was followed by an extensive UK tour. This momentum continued with the highly anticipated *How Would I Know* EP.

Guy's career soon expanded internationally, performing throughout the United States and the Caribbean as a show drummer and theatre musician aboard cruise ships. When the global pandemic brought touring to a halt, Guy returned to the UK and transformed his focus, building a professional home recording studio and establishing himself as a sought-after remote session drummer, recording for a wide range of independent artists and genres.

In recent years, Guy completed a Master's Degree in Learning and Teaching and founded the British Drumming Academy, a global platform dedicated to delivering high-level drumming education to students worldwide. He has also toured extensively across Europe with the acclaimed progressive rock band *Also Eden*. Alongside his performing and recording career, Guy also works as a university lecturer in drums, music performance and music technology and also offers private tuition through the British Drumming Academy, both from his studio and online, continuing to shape the future of drumming through performance, education, and mentorship.

Notable musical projects and credits

Also Eden – UK & European tours and upcoming album/EP releases
Rock The Night – UK's notable live and theatre Rock Show
Shannon Hynes – nominated Country Artist of The Year
Dirt Road Diary – Country Band of The Year
The Second Sons – live drummer for European tours
America Rocks – theatre show
Oli Barton – album, EP and single releases through Coke & Dagger Records
International online session drummer

How to use this book

Who this book is for (in no particular order)

- Drummers of all levels and ages
- Producers
- Beatmakers
- Composers
- Arrangers
- Educators

This book is designed to give you an understanding of how different drum grooves are created in and for different genres so that you can gain a deeper understanding of how to create drums for your own musical situation or if you are playing/producing for someone else.

A large part of what *Drum Groove Creation* aims to achieve is bringing an awareness of the cultural important of these drum grooves to the forefront of musicians' and producers' minds. Whilst some of this information is accessible through other works or by listening to the recordings, this publication will dissect some of the background and key characteristics that acknowledge the creation of these grooves. It is designed to help aspiring drummers of beginner, intermediate and advanced levels. In addition, it is designed to help producers and beatmakers learn how to programme drum grooves and embed this skill into their music creation.

A large part of what *Drum Groove Creation* aims to achieve is bringing an awareness of the cultural importance of these drum grooves to the forefront of the minds of musicians and producers. Whilst some of this information is accessible through other works or listening to the recordings, this publication will dissect some of the background and key characteristics that acknowledge the creation of these grooves. It is designed to help aspiring drummers of beginner, intermediate and advanced levels. In addition, it is designed to help producers and beatmakers learn how to programme drum grooves and embed this skill into their music creation.

Educational settings

This book may also be used for educational purposes, whether in a one-to-one session between an educator and a student or in a group setting. It can be used:

- In a class of drummers
- In a music technology lesson setting

How to use this book

- By producers/beatmakers at home
- By songwriters who want to create drums for their music
- By producers/beatmakers or drummers in one-to-one lessons with their peripatetic tutors
- By drummers learning on their own

Reader benefits

This book gives you an introduction to drum grooves from different genres as a starting point to explore new genres that maybe you haven't listened to or come across in the past.

A greater overarching focus of this book is to showcase the impact of different races and cultures that have contributed to the success of music since the early to mid-20th century, as well as being from different places around the globe. This will hopefully help raise awareness and appreciation for some of these drummers and artists.

As briefly discussed above, this book discusses the ability not only to regurgitate grooves of the past, but also to give musicians and producers the power to create new grooves as a drummer or through programming on music software.

For drummers

Starter, intermediate or advanced?

I don't want to get too tied up in terminology or to define a person's drumming ability as being at a specific level as I feel it can be demoralising and make people feel that they can't push through into the next barrier (you always can!).

However, there are a variety of grooves featured in this book, and they are at different challenge levels. This is not to say, however, that just because you may consider yourself an advanced player you should not learn to play/refine the more simplistic grooves.

Sometimes, it's not what we play but how we play it.

The same goes for beatmakers. It's not always about using or creating the most complex groove; rather, it is about finding the groove that is best for the song. Eventually, you should be able to play all the grooves in this book. However, instead of just going through page by page and struggling, you may decide to work on the simplest grooves first, then the more intermediate examples and, finally, the advanced ones. This can stop you from becoming stuck, getting demoralised and feeling down about your drumming. Always remember: playing the drums is about having fun and enjoying the instrument. Otherwise, what's the point?

The same goes for beatmakers.

How can you use this book?

Reading a groove from this book and playing it a few times, then moving on, isn't enough. Start each groove slowly, then practice to a metronome and gradually increase the rate of speed until you reach your goal tempo. Throughout the book, you will find the following tables, 'Journal Tables' (below), make sure to fill this in and use it to help aid your practice and your musical journey.

Journal table example

When	What	Tempo	Notes/Reflections

How to use this book

The point of these journal tables is to keep track of your progress and make a note of challenges, interesting ideas or things you have discovered.

Also, as mentioned above, it can sometimes be easy to feel crushed by the challenges of playing an instrument. These journal tables can be used for you to look back on to see how much progress you have made from when you first started using *Drum Groove Creation*.

The purpose of these journal tables is to keep track of your progress and make a note of challenges, interesting ideas or things you have discovered.

Also, as mentioned above, it can sometimes be easy to feel crushed by the challenges of playing an instrument and these journal tables can be used for you to look back on to see how much progress you have made from when you first started using *Drum Groove Creation*.

Working with a metronome

I will go out on a limb and say that most drummers know that we should all be practicing to a metronome. However, whilst I strongly encourage everyone to do this, there is one small problem. This can simply make you proficient at playing to a metronome, but some drummers may become reliant on this. Once it's taken away, their timing crumbles because they haven't developed a sense of internal time. Because of this, there are a few exercises that we can try to start to develop this sense of internal time and to become less reliant on a metronome.

1. Play with the metronome on off-beats.
2. Play with the metronome on two and four.
3. Play only with the first beat in the bar.
4. Play with one bar of click and then a bar without.

If you do the above four exercises with every groove in this book (plus the ones you create yourself), I guarantee that your sense of time will improve hugely.

For beatmakers/producers

How to use this book

If you are a beatmaker or producer who is looking to learn how to programme and create your own drum grooves, then this is the book for you. You don't need to concern yourslef with the drum notation (unless you want to); instead, just look at the grid below the drum set notation. These grooves can be programmed into your Digital Audio Workstation (DAW) and manipulated as you see fit. What it will do is give you a grounding in that genre and allow you to build and grow from it. These grooves can then be used in alongside the chapters on *Groove Amalgamation* and *Groove Creation* with the *Diamond Cutter Method*.

Listening

This is applicable to everyone. One of the best ways to get to understand different genres is to listen to them. This could be through finding a playlist of that genre online or by googling the artists best known in that style. That will lead you from one to another and give you more ideas than are listed in this book.

When listening, and engaging in active listening, it is important to pay attention not only to the drums, but also the whole band/artist. This is how we begin not only to understand the drum parts, but also how they fit within the music. It's all about context, and how the right drum parts work for and complement the music in question.

Limbs

For drummers, we often think in terms of limbs and how we use them on the drums. For beatmakers, this is not necessarily at the forefront of our minds. However, in terms of referral, I will sometimes name specific limbs. This is because these limbs can be orchestrated onto different percussive surfaces. This is for the purposes of writing and reference, so, beatmakers, it's still quite a good way to think about it so your beats can sound realistic and musical. Consider the four limbs (most) drummers would have; use this as a foundation and then build on it as you wish.

For definitions of many of the terms found within this book, please consult the *Glossary* section at the end of this book. However, before we kick off Chapter 1, I wanted to explore two terms...

A *groove* may be thought of as an established, repeated beat on the drum set. This is a repeated phrase that helps set the rhythm and tempo/timing. This is often the core of what drums will do in a song to help the music around it. It lays a clear foundation for the other instruments to build on.

Feel

Feel may be considered a drummer's/producer's/beatmaker's unique stamp and personality within a groove. It can reflect the way they subtly emphasise certain notes, vary dynamics/velocity, and play with the rhythms to individual parts of the drum set to create a new sonic experience for a listener.

What elements go into achieving feel?

Time

Maintaining a steady sense of time is a cornerstone for good feel. It helps create a hypnotic, unrelenting experience to a track.

Appropriate sound and dynamics

Feel also involves using appropriate sounds for the style/genre. In this book, this is explored through the sound substitution concept. Varying the volume and dynamics between individual drums can add to the intended expression and flavour.

Accents and non-accents

Following on from the different dynamics between instruments, utilising a combination of accents and non-accents within grooves is also key to achieving a powerful feel.

Individual expression

Feel can allow us to express individuality and personality through our grooves, and to create our own unique sound. It can help bring a song's energy to life through timing, appropriate sounds, dynamics and expression.

How to use this book

The power of feel

Given how we have defined feel above, it is clear that this aspect of drumming has enormous power. It is feel that makes people want to dance, that helps evoke the emotions of the song and that ultimately helps encapsulate and project the intended sonic energy of what we are trying to achieve.

Can you learn feel, and programme it – or are you born with it?

Feel does not have to be thought of in black and white terms. I fundamentally disagree with the mentality of – 'either you've got it...or you haven't'. The tools within this book will help you create drum grooves with a power feel to help to take your playing, programming and music to the next level.

British Drumming Academy

Using the online Groove Creator tool

As part of this book, we have designed an online Groove Creator tool through the British Drumming Academy that was founded in 2024.

The British Drumming Academy's online groove creator tool can be used to give audio examples of the grooves in this book (and the ways to teaches you to create them). In a similar way to programming MIDI, you can select your time signature, subdivision and the parts of the drum set you want to use. From there, click and add your grooves. These can also be shown in notation and used to check against the grooves in the book so you can hear how they sound.

Visit britishdrummingacademy.com to sign up for your seven-day FREE TRIAL to use our online groove creator tool.

British Drumming Academy

Introduction

Welcome to *Drum Groove Creation: For Creative Drummers and Producers*

This book is designed to equip the modern drummer, producer and beatmaker with the knowledge of notable and genre-defining drum grooves from the 20th and 21st centuries, as well as an initial insight into some of the rhythms used within different cultures outside of (what may be considered) mainstream music. From acquiring this knowledge, you will learn the tools to be able to create your own drum grooves, either for yourself or for different musical projects. *Drum Groove Creation* will take you through the absolute essentials of groove construction, which are the tools you need. They are based on defining characteristics from the 20th century.

Having worked through this book, you will be ready to play, programme and create drum grooves in a variety of musical styles; no matter the scenario, you can build what is needed. From acquiring this knowledge, you will learn the tools to be able to create your own drum grooves, either for yourself or for different musical projects.

So, get strapped in and enjoy the journey as we explore *Drum Groove Creation: For Creative Drummers and Producers*

In my role as a professional drummer and educator, I often find that drummers want to learn how to play different styles. Equally, I often encounter a lot of music creators who are beginning to explore music technology in order to create their own music and who want to learn how to programme and create drumbeats for themselves, instead of using prearranged loops.

The downside of using pre-made drum grooves (and not altering or editing them for your own means) is that your music runs the risk of sounding identical to everyone else's. I'm sure that's not what you want; every artist wants to sound unique and to express their individual musical personality. There's nothing worse than creating a great track that showcases you as an artist and then adding a generic drum loop that doesn't quite fit what you had imagined. This book will help you overcome that obstacle.

This is not to take anything away from material that is out there already, but growing up and learning drums I used to feel that many drum books out there primarily focused on memorisation, regurgitation and the ability to read notation. However, in this book I would like to present this information in the forms of both drum notation and in a piano roll-style format (as it would appear in a DAW if you were programming MIDI).

Turning now to those drummers, producers and beatmakers who do not typically read (or struggle with) drum set notation, the goal behind this book is to give you the blueprint so you that you might build the house. By this I mean that it contains a series of grooves, tools and musical characteristics, which you must take away and make your own. This is through variation and manipulation as shown in the Groove Creation and Groove Amalgamation chapters and using our Diamond Cutter Method (more on this later).

This helps make the material accessible to drummers, producers and beatmakers who do not typically read (or struggle with) drum set notation. It is also useful for producers, when they begin to work in musical situations and do not wish to simply replicate verbatim what has come before. *Drum Groove Creation* will show you how to amalgamate styles, characteristics, and concepts and to create new and distinctive drum grooves.

Chapter 1 Starting essentials

Welcome to the first chapter of this book: Starting essentials. This chapter aims to give you an understanding of what we need to get started on our Drum Groove Creation journey.

Essentials: Understanding subdivisions

Subdivision is essentially how we divide the beats of the bar equally into different note types: quarters, 8th notes, 8th note triplets, 16th notes, 16th note triplets and 32nd notes.

A bar of 4/4 can be divided into four quarter notes (quite literally what 4/4 means), this can then get equally divided into eight 8th notes, sixteen 16th notes and so on. One way to think about subdivision is that it refers to how many notes fit inside of the pulse (each quarter note). Please see the 'Rhythm Tree' below.

DOI: 10.4324/9781003585169-1

Rhythm Tree

Counted as

1 · e · & · a ·	2 · e · & · a ·

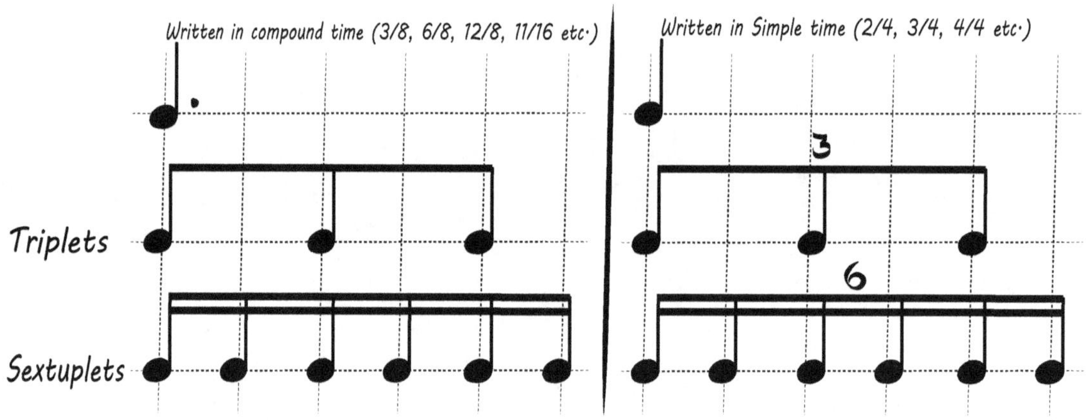

Counted as

1 & a	2 & a

Or

1 tri plet & tri plet	2 tri plet & tri plet

A note on subdivisions

There are a few other note types of which it is worth being aware of. These include whole notes, a note that lasts for the whole bar (this may be a cymbal hit that sustains its sound), and half notes, which is half as long, lasting for half a bar. Equally, there are some subdivisions that this book does not really dive into: quintuplets, septuplets and nonuplets. These are essentially when the pulse gets split into even sets of five, seven and nine.

Essentials: Notation and piano roll grid

Generally, the way I designed this book was, drummers, read the notation, and beatmakers/producers, use the grid so you know what to programme. But it's up to you; learn to understand both if you wish and do whatever works for you. The journey may be different, but the destination is the same.

In this image, I show you where each part of the drum set is represented on the stave.

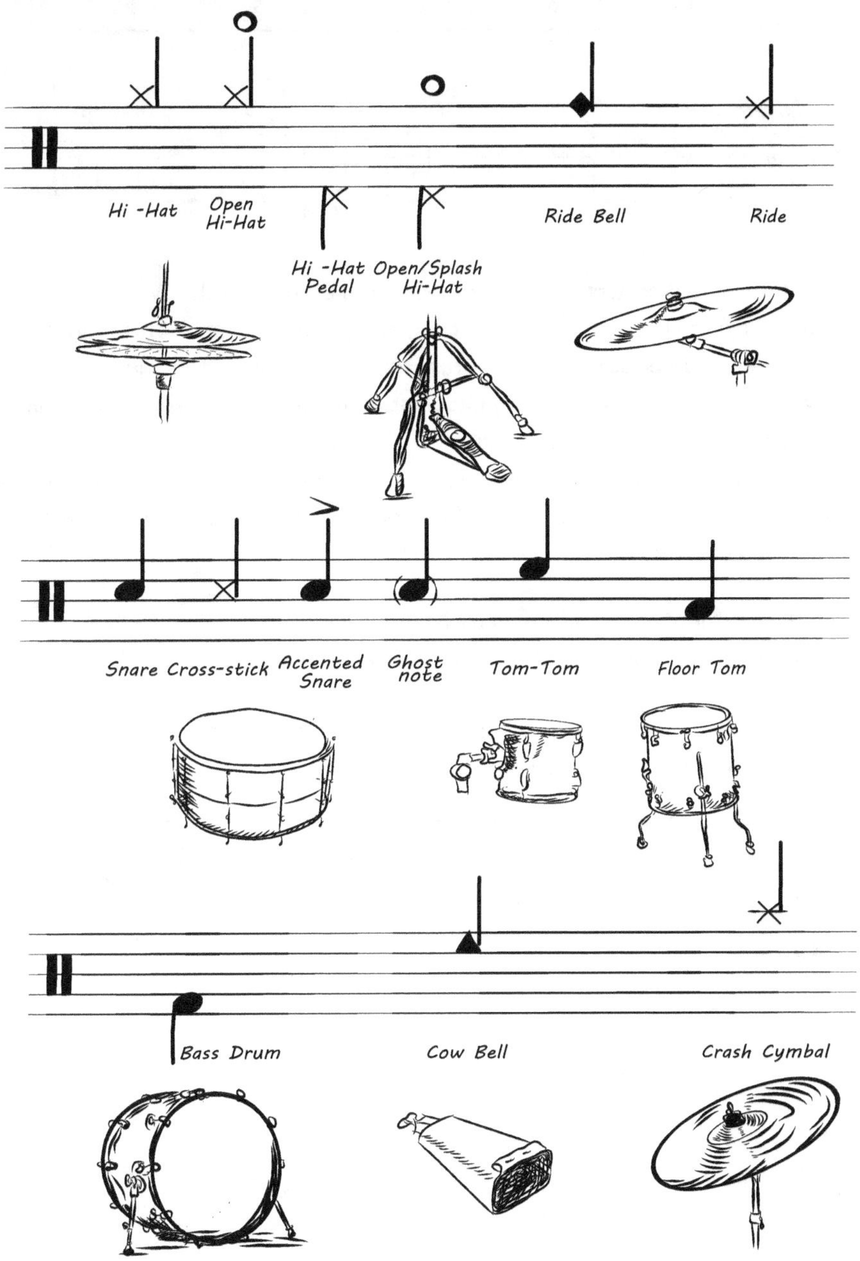

Drum Groove Creation

The grid explained

In a similar way, we have the Piano Roll Grid. This is typically what most beatmakers and producers will use in a digital audio workstation (DAW) when creating Drum Grooves. This displays the drum/cymbal to use in your DAW and the subdivision of the groove we are creating.

Groove 1: The 'Money Beat'

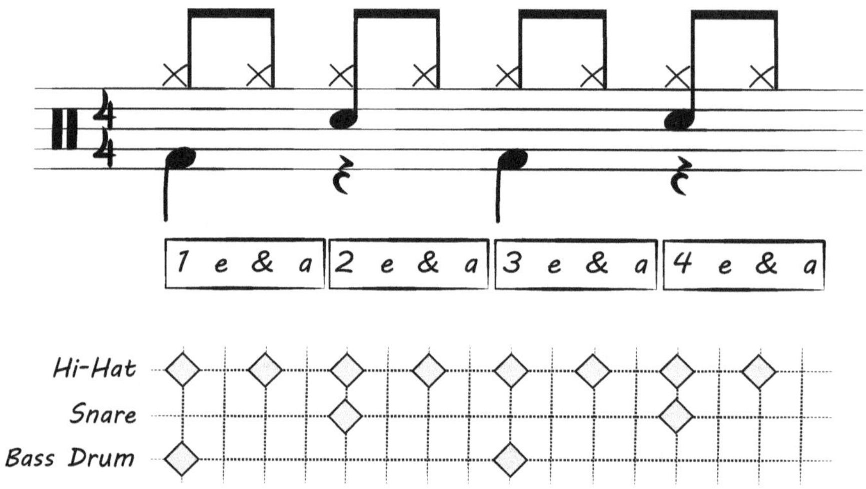

Arguably, the most important 'modern' groove for the drums is the 'Money Beat'. This is aptly named because it's what helps you keep your gigs and get paid. This groove works at very slow tempos as well as faster ones. Drummers, this in and of itself may take some time to achieve comfortably. **No-one should underestimate the effectiveness of simplicity.** This beat is heard on literally thousands of songs for a reason.

The Money Beat may be viewed as having some of the characteristics and cornerstones for many of the grooves in this book, and for Groove Creation in general. Here are some of the staple characteristics to be aware of and to take forward:

- The Hi-Hats (or often ride/crash) plays 8th notes.
- The snare is on beats '2 and 4' and is called the **backbeat.**

These characteristics, of Hi-Hats playing 8th notes and snare drum on '2 and 4', is what we will call **ostinato one.** Note how this is used in Groove 1 in the example shown.

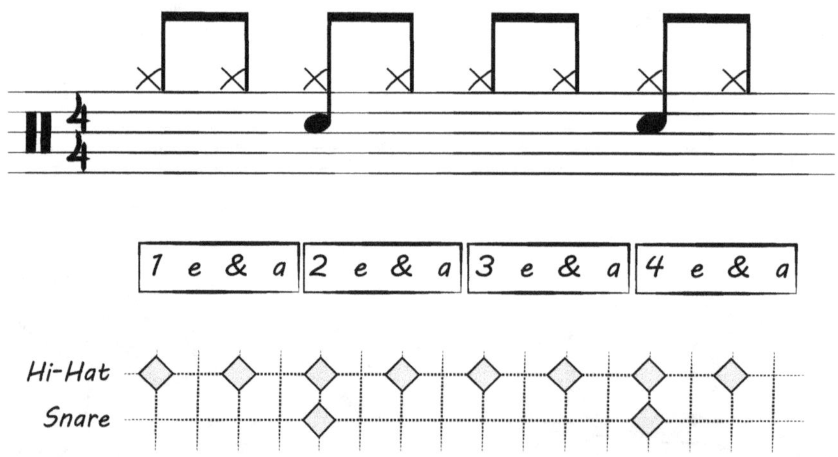

Groove 2: The 'Money Beat' variation

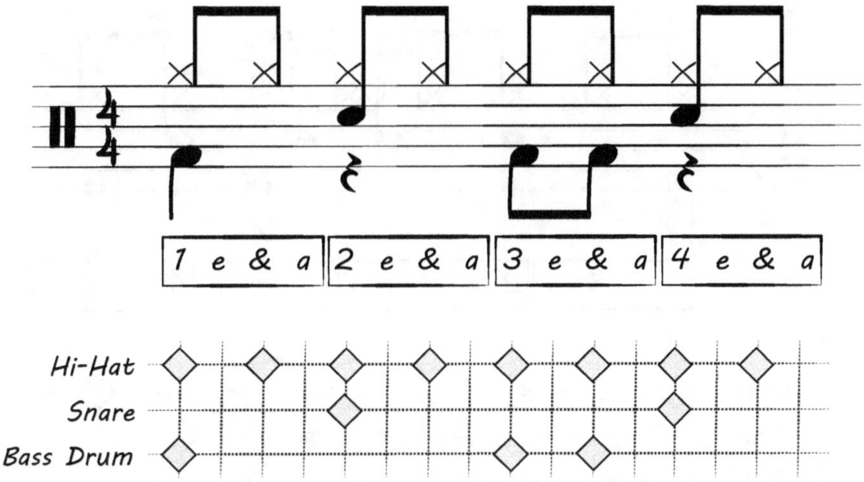

That brings me to my next point in terms of the hi-hats. The majority of the grooves within this book are written on the hi-hats. However, any groove can be moved to the ride cymbal, and vice versa. But note that any grooves with open and closed hi-hat notes may alter slightly. Equally, if you wanted to take it one step further, the hi-hat could not only be replaced with the ride cymbal but just as easily substituted with the floor tom, crash cymbal, rim or side of the drum or a piece of handheld percussion like a shaker or a jingle stick. The list goes on. Push this, be creative and explore sound substitution. There will be more on this and how to use it later in the book.

We could literally fill an entire book on variations of the 'Money Beat'. Arguably, a lot of these grooves in some way derive from it (note the persistent presence of **ostinato one**). Groove 2 keeps the groove-building characteristics with the hi-hats on 8th notes and the snare playing the Backbeat. What we begin to see here is our first variation in the bass drum pattern.

Groove 3: 16th notes on the hi-hats

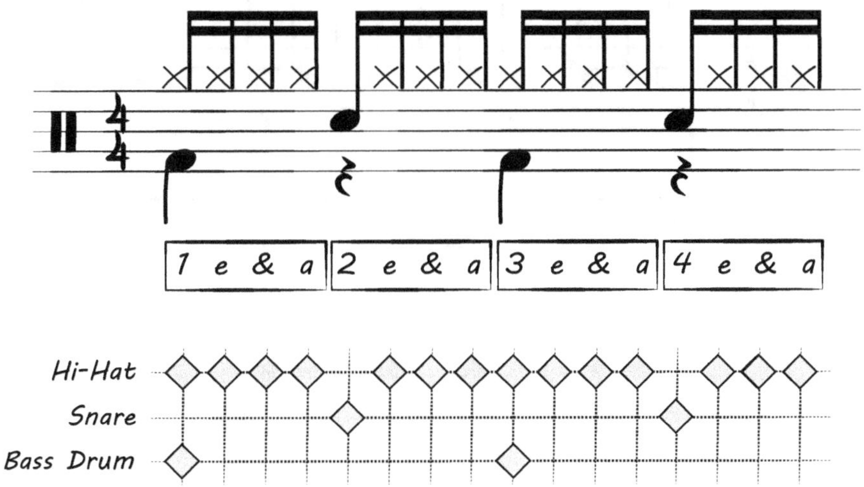

Our third groove utilizes 16th notes on the hi-hats. This groove lends itself to many different genres with different variations in the bass drum. This can be a great way to give a song a sense of drive and urgency. Equally, at a slower tempo it may provide a laid-back feel; it all depends on the context of the song.

Drummers: you may choose to play this one- or two-handed depending on speed and preference.

Groove 4: Ghost notes example one

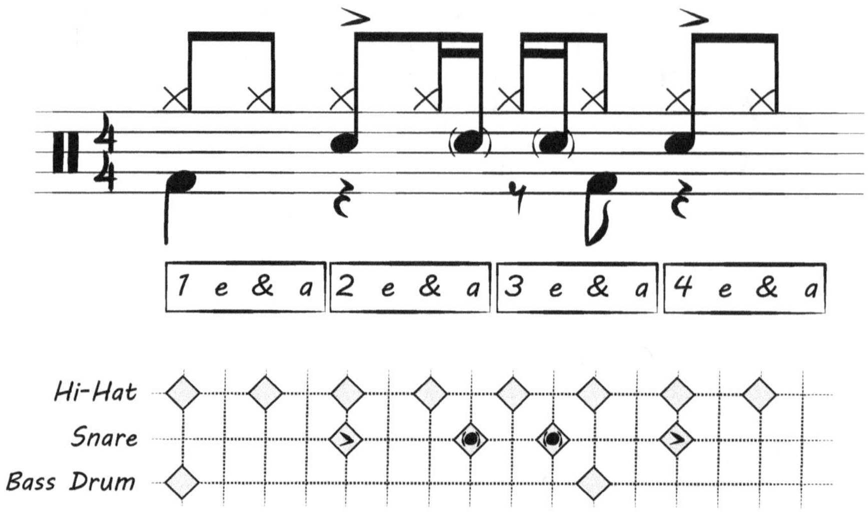

Ghost notes or grace notes: these are different words for the same texture that we are trying to achieve. Essentially, these are unaccented notes that are quieter in comparison to the backbeat. Note that the 8th notes in the hi-hats are still present, but now we are venturing into the subdivision of 16th notes on the snare drum.

Beatmakers: make sure you adjust your velocities here otherwise all your snare drum notes will be at the same volume, and your groove will lack texture and dynamics.

Keep the ghost notes quieter than the backbeat.

Groove 5: Ghost notes example two

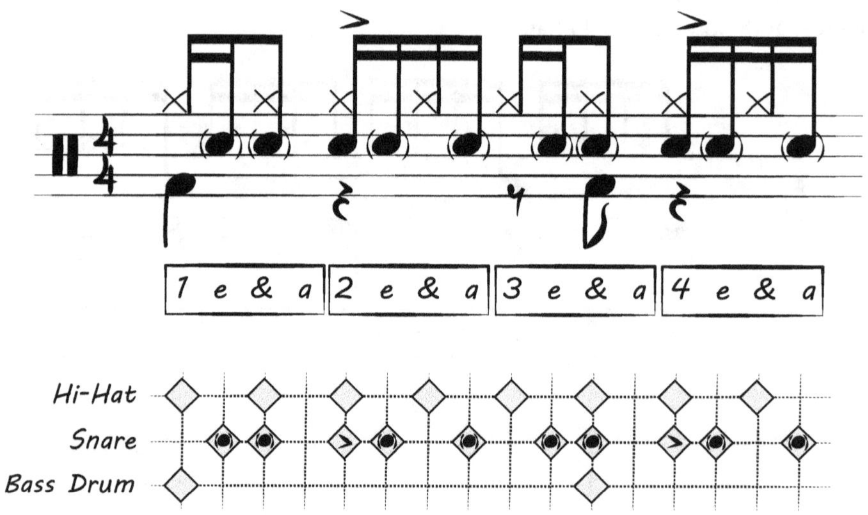

Slightly trickier and more complex here, but still a variation from the previous groove. You should note the similarities. Whilst both grooves use ghost notes, you will achieve a slightly different sense of texture and feel through this variation.

Groove 6: Dance beat example one

As mentioned in Groove 1, the 'Money Beat' is arguably the cornerstone of modern grooves and forms the foundation for many grooves in this book. Groove 6 showcases that. At first glance, it may look quite different, but we are still retaining our backbeat. However, we have now added a bass drum on all four beats of the bar (this is commonly referred to as four on the floor) and the hi-hats are playing on the off-beats. This groove is often found in Disco, House, Dance and Pop music. This is a key groove to be aware of for any beatmaker/DJ who wants their audience to dance.

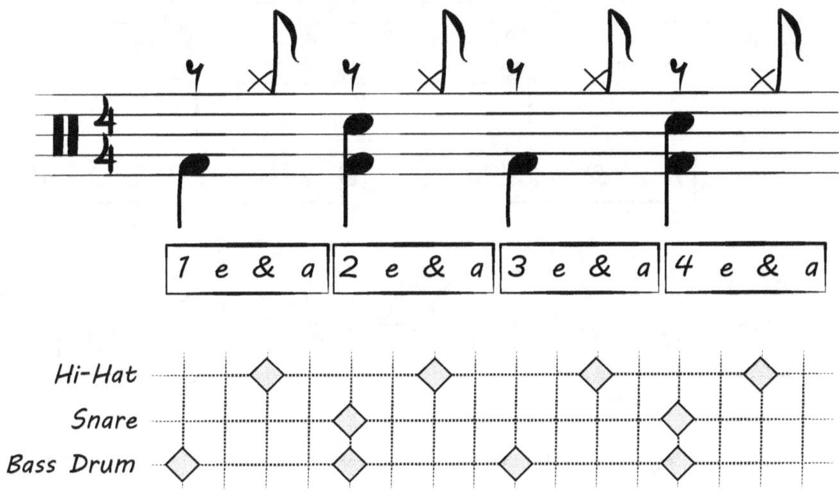

Another perspective?

Another way to view this groove is that we are still thinking in the way of 8th notes on the hi-hats, but we should now think of them in 'groups of two' and we are just removing the first note from each of the pairs. Essentially, removing all the 8th notes on the beat, but retaining the ones off the beat.

Groove 7: Dance beat example two

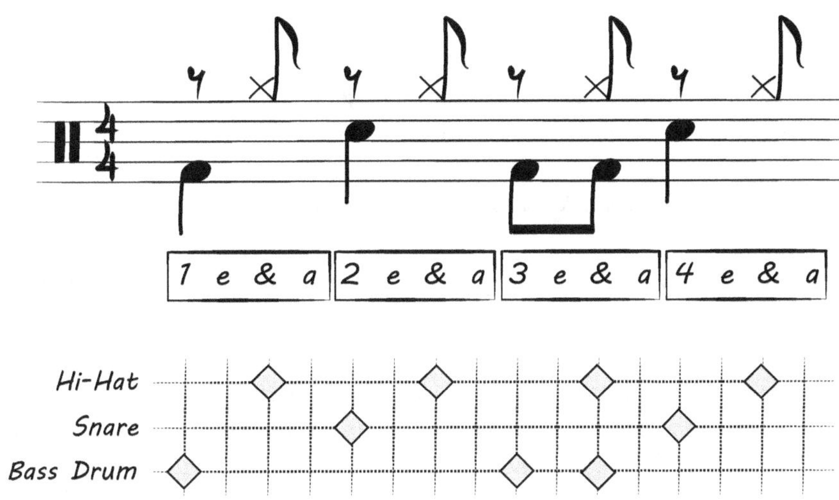

Groove 7 is a variation on Groove 6. Here, we are keeping the off-beat hi-hat feel but varying the bass drum slightly. It is worth noting that with both of these dance-style grooves, the hi-hats could both be played closed or also with the opening/close effect. This will give you a variation of feel and could alter through the course of a track to facilitate different intensities at different musical points. To hear this in action, listen to more or less any disco track from the 1970s.

Groove 8: Combination groove

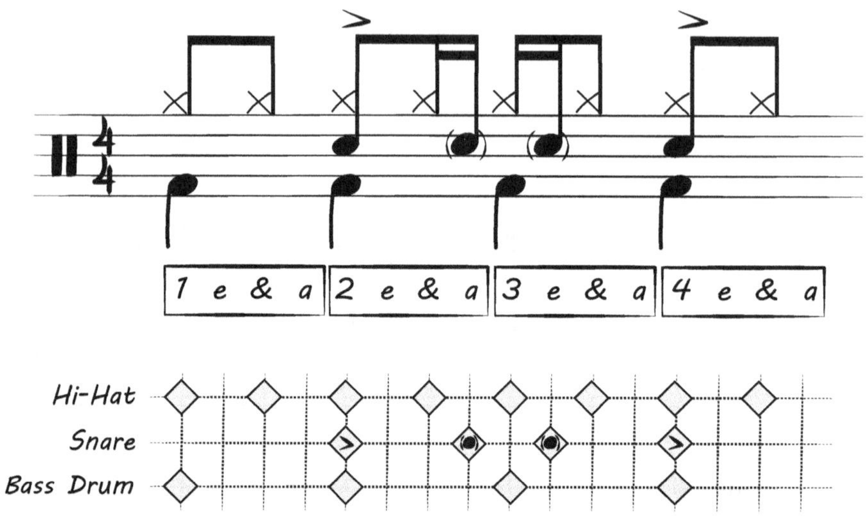

So, this is where we begin our first step of our journey into Groove Creation to see what we can cultivate at this point. What we have here is a bit of a sneak peek of a later chapter: *Groove amalgamation*. This (explained in more detail later in the book) is when we are combining different grooves shown in this book to create something new for ourselves and our music. Whilst this groove may not be something that has never been heard before, it is more about the journey of how we have come to it by using this book. Let me explain...

Here are the elements we have taken from some of the grooves we have encountered so far to be able to create something new.

Hi-hats – 8th notes as shown in Groove 1 – The 'Money Beat'
Snare drum

- taken from Groove 4 when we got introduced to ghost notes
- four to the floor, taken from Groove 6, our first Dance Groove

Bass drum – four to the floor, taken from Groove 6, our first Dance Groove.

From combining these different elements together, we get Groove 8, the Combination Groove.

16th note bass drum placements & the importance within groove creation
This is not a new concept designed by me. Drummers and educators such as Gary Chaffee, as well as Benny Greb in his wonderful book, *The Language of Drumming*, have helped bring this to the masses. Our goal is seeing how we can use this to expand our grooves and create new ones with it.

In essence, in a bar of 4/4 we have, sixteen 16th notes (makes sense, right?), so that means that there is an opportunity for us to place bass drum notes anywhere in this bar on any of these 16 placements available. This would not just limit us to one note per bar; it suggests, rather, that a single note could be placed anywhere multiple notes in any combination as well as all 16 notes together. Here is a blank example, but with ostinato one written in.

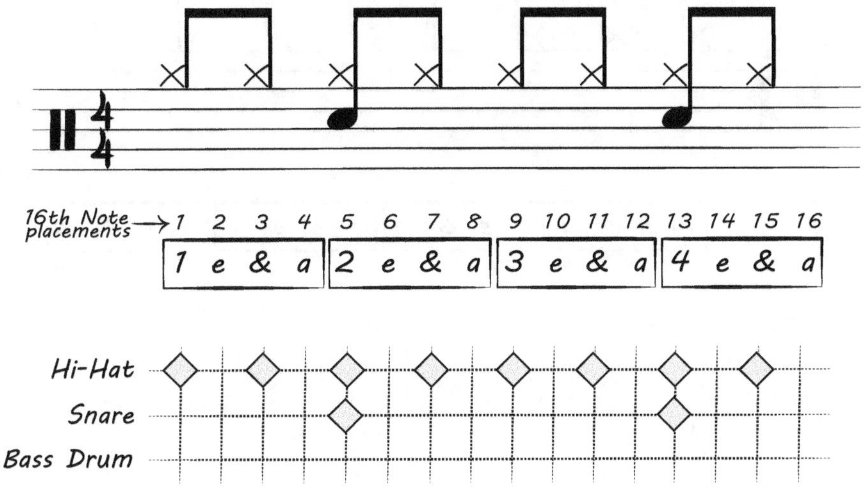

Note grid and placements

Here is a figure simply showing how these would work in relation to ostinato one and the different places we could put our bass drum notes. Note the numbers 1–16. Our bass drum may be played on none, any or all of these (in any order) to create different bass drum patterns. In this example we have placed bass drum notes on numbers, 1, 6, 10 and 16.

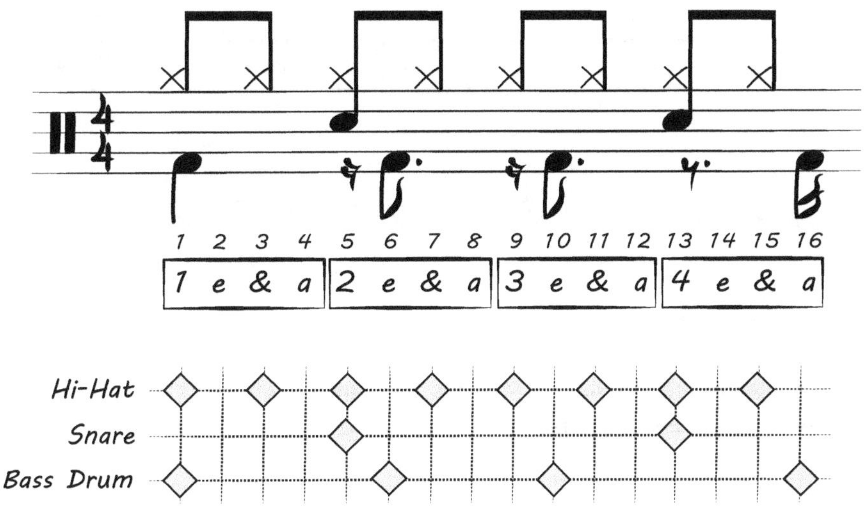

The hi-hats

Now it's time to discuss some of the different ways that we can approach the hi-hats, something that is often overlooked but which is one of the keys to unlocking great feel on the drums – and consequently giving feel to a whole song. Below are some different hi-hat ostinatos that may be used over the top of grooves. These can be applied to the above melodies to create different sounds and feels. We will also revisit this concept later in the chapter on Groove Creation.

Drum Groove Creation

Option 1 – Straight 8th notes

This one is great for driving a song and giving it a sense of forward momentum.

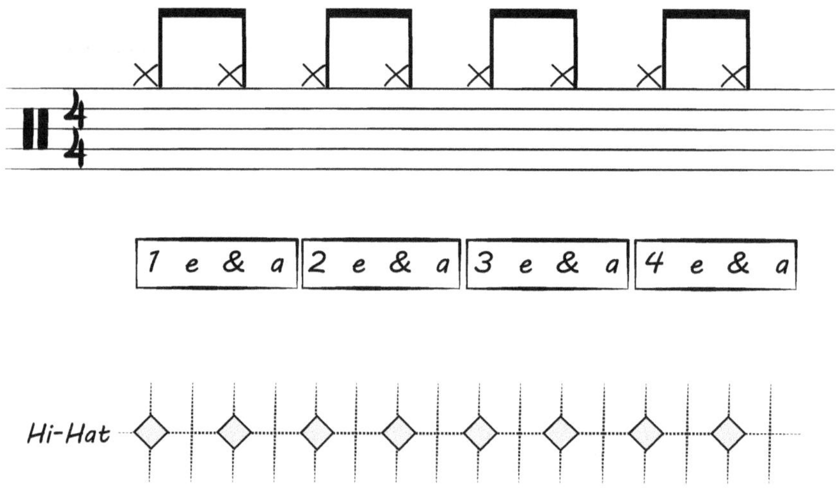

Option 2 – Digging in

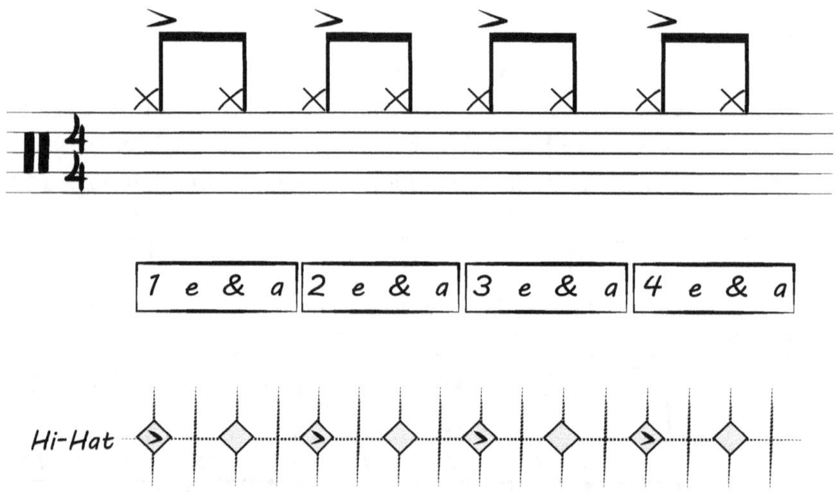

Option 3 – Upbeat feel

This is the one that makes people want to dance. Sometimes at faster tempos I may start to drop the down beats and just keep the upbeat hi-hats.

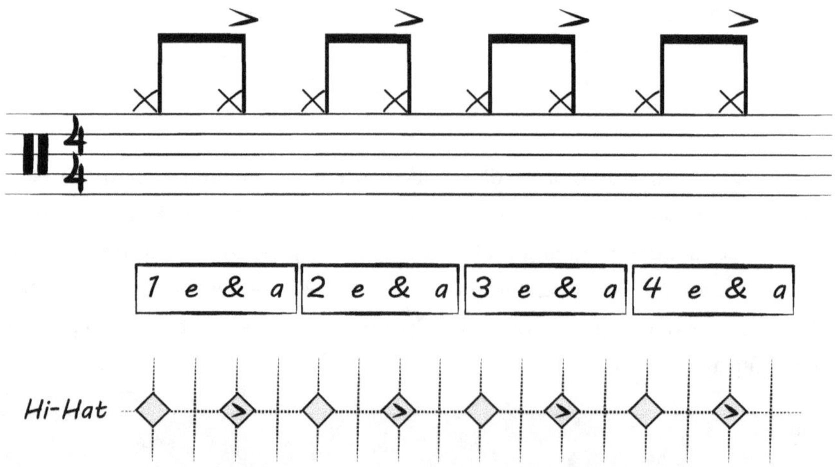

Groove creation: Challenge one

This book features groove creation challenges at the end of the chapters. For this chapter, follow the follow instructions to create your first set of drum grooves.

Step 1 – select a hi-hat option from the list shown.
Step 2 – select a snare drum pattern from the examples in this chapter.
Step 3 – this is the key point. Select any numbers between 1 and 16 in relation to the bass drum. This is where we will put our bass drum notes to create new grooves.

Combining these different hi-hat options, different snare patterns and then selecting different numbers for the bass drum gives you a roadmap for creating hundreds of useable drum grooves. Here is an example to get you started.

Groove creation challenge one example

Step 1 – Hi-hat option two – digging in.
Step 2 – snare drum pattern from groove four.
Step 3 – using our 16th note grid, numbers 1, 3 and 14.

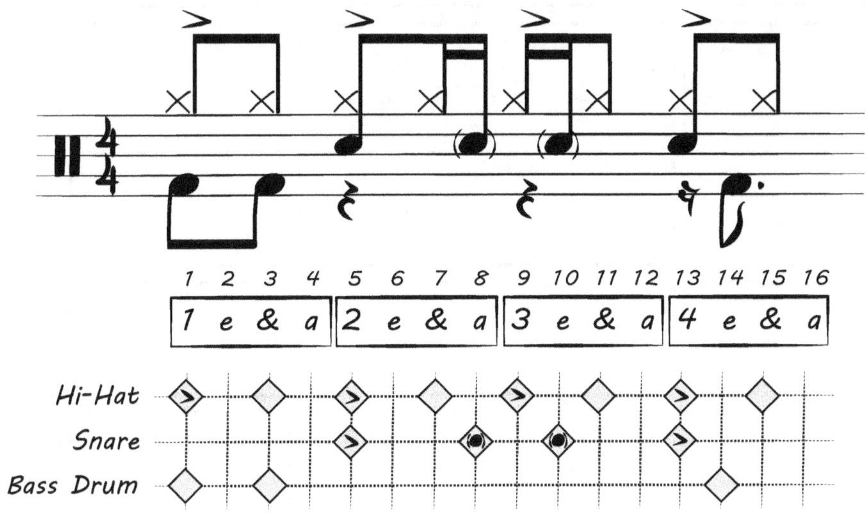

Drum Groove Creation

Journal table

Remember to make note and keep track of your progress by using the journal table.

When	What	Tempo	Notes/Reflections

Chapter 2 The Blues and Shuffles

Whilst this book is not designed to be an in-depth history lesson into each of the musical styles and genres involved, I feel that it is important to acknowledge both the origins of each style and some of the key artists and drummers responsible for having a significant impact.

One of the darkest episodes in human history was the transatlantic slave trade when it is estimated that between 10 and 12 million Africans were captured and transported by slavers over the Atlantic Ocean to the Americas on a journey that could last up to several months. Slaves were chained together in abhorrent conditions, and many died over the course of the voyage.

The slave trade was one of the main features of American life until the official abolition of slavery in the 19th century. As history progressed, slaves began to obtain their freedom or run away from plantations. One of the more enlightened areas consequently occupied as a kind of musical haven was the city of New Orleans. Every Sunday, freed people gathered in Congo Square to play their music together and dance. This city was (and still is) brimming with Spanish and French influence, and is often referred to as the birthplace of Jazz.

I want to take a moment and talk about an important aspect of New Orleans music: Second Line drumming. New Orleans is a city of parties, colour and celebration; even the funerals can be huge affairs with grand street parades. The first line of a parade often features dancers and brass players. Behind this are the drummers, armed with either a snare drum or a bass drum. New Orleans music is known for having a great in-between straight and swung feel to it. Capturing this feel can be challenging; it is something quite unique in the world of music.

Genre overview

There is room for deliberation into the origins of Blues music. Fundamentally, it was influenced by the African slaves working on plantation fields, and the songs that came from there. The expressive vocal form and the lyrical narrative was often centred on the subject of oppression. This initial form of Blues music largely derives from the experiences of Black agricultural workers in the southern states of the United States of America. These can be traced largely to Georgia, the Carolinas, Texas and Mississippi.

Later in the 20th century, rural Blues encountered the technological advancements of electric guitars and amplifiers. This meant that drummers playing the modern drum set (as opposed to handheld percussion, 1940s style drums or clapping) were able to play as part of the ensemble as the volumes could be matched. This laid a path for some of the Shuffle grooves we will encounter below and the variations evolved from these.

DOI:10.4324/9781003585169-2

Drum Groove Creation

Key artists to listen to

Bessie Smith
Ma Rainey
Robert Johnson
Muddy Waters
BB King
Blind Lemon Jefferson
Charley Patton

Key drummers

Elgin Evans
Fred Below
Odie Payne

It is worth noting that, due to the age of some of these recordings and the lack of certified information, it can be difficult to tell exactly who appears on any individual recording. Many of these drummers were session players who would have been playing with the same artists.

Groove 9: The 12/8

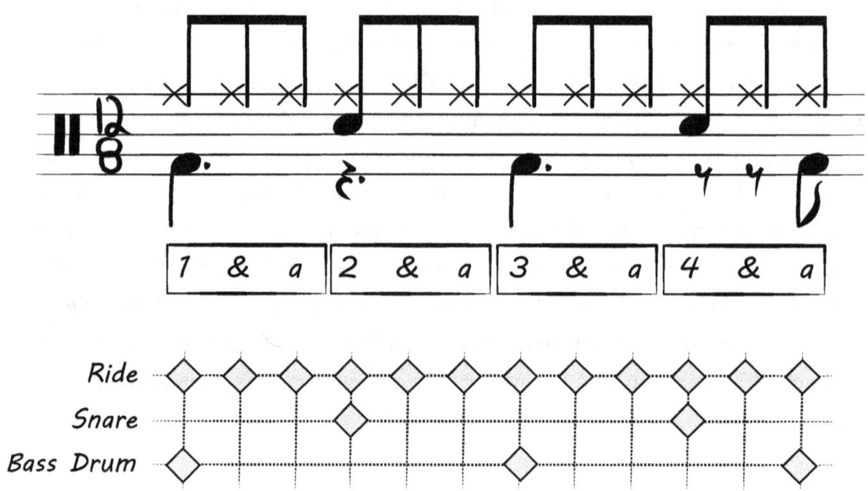

So, this is the first groove that we have encountered that is not written in 4/4. Thus, what exactly does that mean? In this scenario we are viewing the groove as having twelve eighth note triplets. I would not recommend counting to 12 each time (this becomes a lot, especially at faster tempos). Rather, let's count it as follows: 1 + a, 2 + a, 3 + a, 4 + a. This means that we are essentially feeling this groove in triplets.

Producers/Beatmakers: remember to alter the subdivision in your DAW to 8th note triplets to be able to programme this.

NB You would sound this out as 'one and ah, two and ah' etc.

Groove 10: The Shuffle

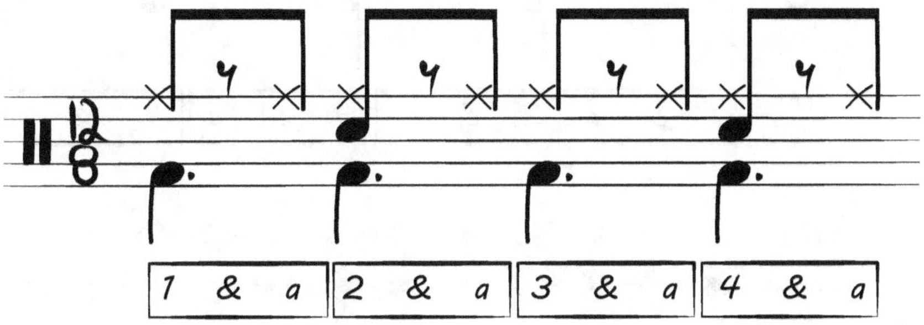

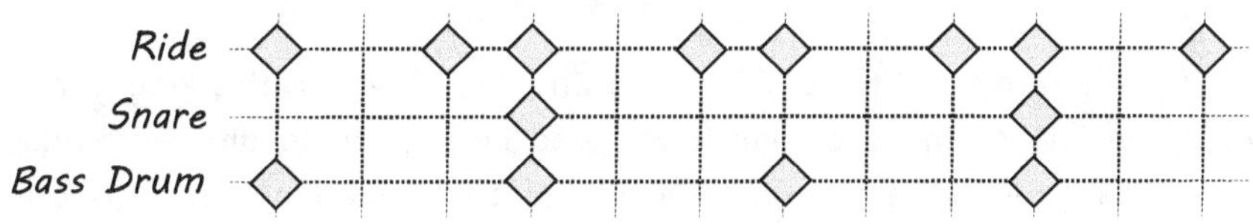

Often overlooked and taken for granted is the Shuffle Groove. So, what exactly is a Shuffle? Well, if we look at our previous example of a 12/8 groove, our ride/hi-hats are playing 8th note triplets. The Shuffle is exactly the same, except we are omitting the middle note of each triplet. *This could be counted as 'one ah, two ah, three ah, four ah'. Essentially just missing out the '+' from the previous example.*

One of the key things to watch out for here is the consistent swing feel in the ride/hi-hats. Often in Blues music, our ride/hi-hat is emulating the rhythm that the guitarist is doing with their strumming hand.

Experiment with the amount of swing that will be available for you to alter in your Digital Audio Workstation (DAW). Use your *'swing-o-meter'* as I like to call it. This is essentially a form of quantisation. Please also note that these Shuffle Grooves would work in a 4/4 context. Just make sure the tempo fits the track you are creating the drums for.

Shuffle variations

There are many variations of the Shuffle Groove, including options with the bass drum, ghost notes and different hi-hat and ride cymbal options. This book looks at some of the key examples that you will need for different musical situations but keep exploring and creating your own variations. For more ideas, explore and listen to performances by the key artists and drummers listed at the start of this chapter.

Groove 11: The Texas Shuffle

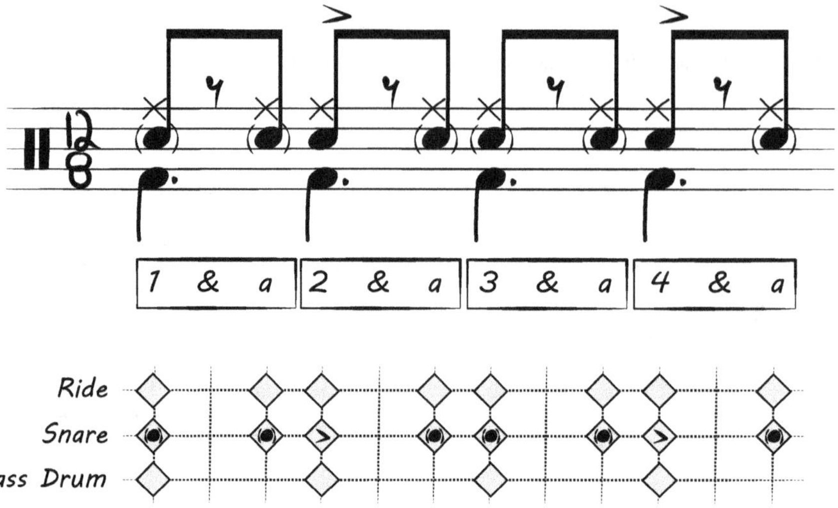

Groove 11 is often referred to as 'The Texas Shuffle'; however, this kind of music is not exclusively played in Texas and it can also be applied to different situations. Personally, I have found that this groove can lend itself quite nicely to a Rock 'n' Roll or Rockabilly song.

In a similar way to the ghost notes we saw earlier in Chapter 1, in your DAW (or snare drum hand), experiment with the contrast between the backbeat and the ghost notes here. Again, unaccented notes should be softer than the accented notes. Adjust your velocity in your DAW accordingly.

Groove 12: The Rock Shuffle

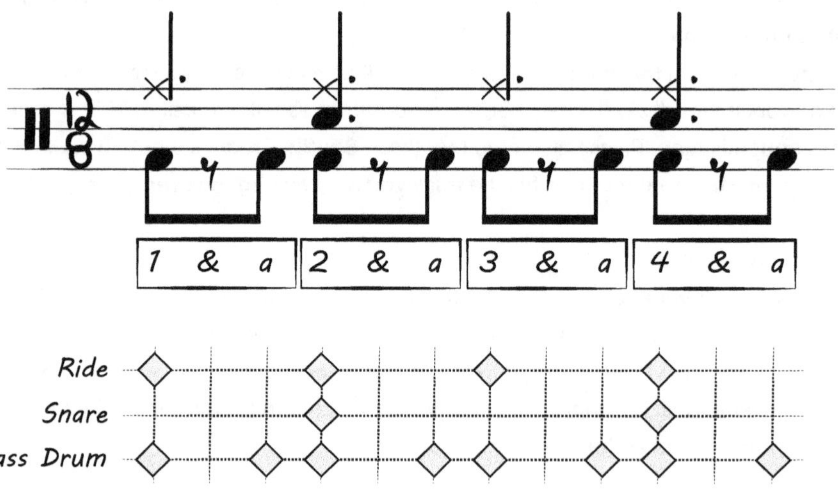

I really love this groove. It's a little heavier alternative to some of the other grooves that we have seen so far. One such example of this being played in context is Rainbow's *'Long Live Rock N Roll'*, recorded with the Late Great 'Cozy' Powell on drums. You can afford to gone heavier on the ride cymbal here in order to match the power of the bass drum. You can afford to gone heavier on the ride cymbal here to match the power of the bass drum.

Groove 13: The Jazz Shuffle

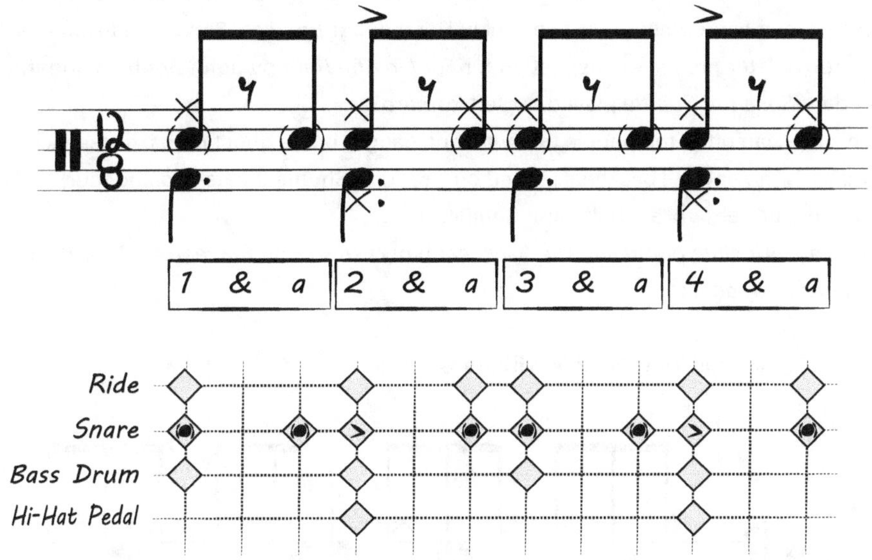

The Jazz Shuffle. I was very tempted to put this groove in our chapter on Jazz that features later in this book. However, it seems a good place to start pushing ourselves a little in this Shuffle section. So, let's break it down. Essentially, on the cymbals we are utilising what is known as a Jazz ride cymbal pattern; the other hand is playing a Shuffle rhythm on the snare, the bass drum on quarter notes and the hi-hat foot on two and four.

Drummers: build this one up one limb at a time; take it slowly to begin with. This is also our first real introduction to some four-way coordination. I also wanted to showcase that whilst some grooves may have the name of a genre in front of them, they can, and 100% should, be used in different contexts.

Groove 14: The Half Time Shuffle

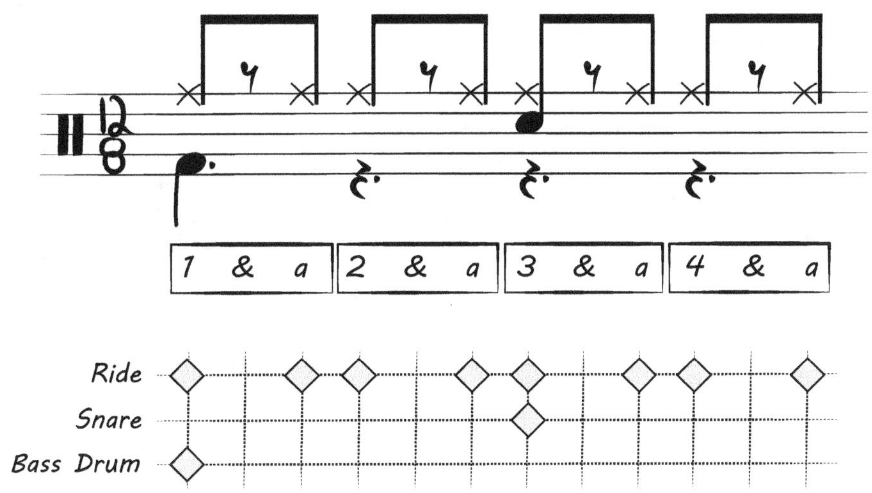

Drum Groove Creation

This is a very popular one groove amongst drummers: the Half Time Shuffle. This particular groove is also referred to as The Purdie Shuffle after its originator, Bernard Purdie. Examples of this groove can be heard on two Steely Dan songs: *Babylon Sisters* and *Home at Last*. This groove later proved an inspiration for *Fool in The Rain* by John Bonham and then the legendary *Rosanna Shuffle* played by Jeff Porcaro.

This one can often trip people up by getting caught in the ghost notes. In essence, all we are doing is moving the backbeats found on a regular Shuffle (beats two and four) and moving it to beat three; voilà a Half Time Shuffle.

NB This principle of moving the backbeat works the same for regular 8th and 16th note-based grooves too.

Groove 15: The Half Time Shuffle with ghost notes

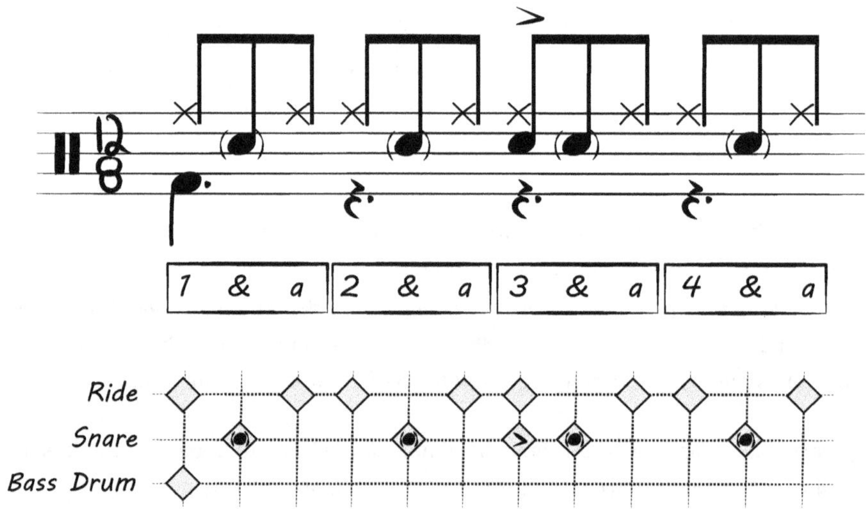

The tricky thing with this one is the ghost notes, so start with everything except these. Once the rest of the groove is sitting comfortably, just gradually add them in one at a time whilst playing at a slow tempo.

Groove 16: The Stepped Shuffle

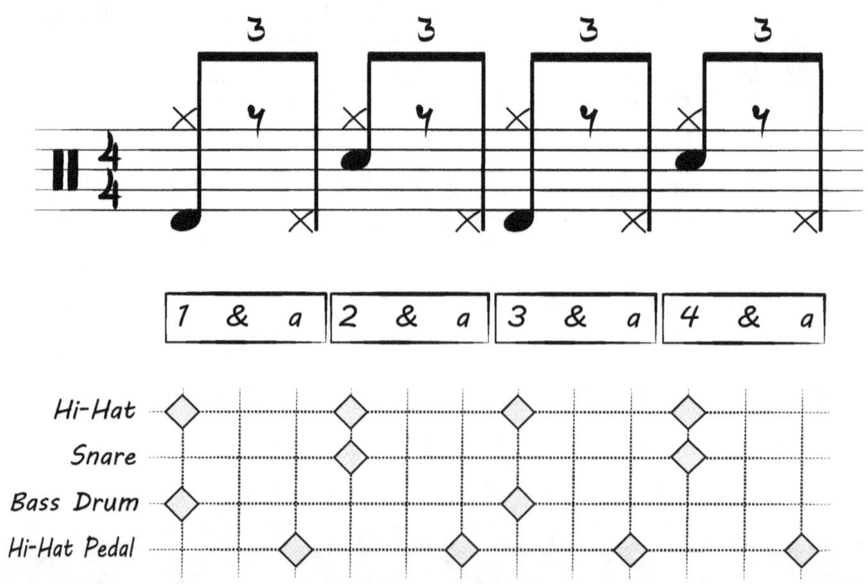

Popularised by the great Steve Gadd, this is a variation to the Shuffles we have seen previously. Instead of everything being played on the hi-hats, in this case every other note is played with the hi-hat pedal instead. Whilst this may seem like a small and insignificant variation, it does provide a subtle, yet distinct feel. Drummers, this opens up a co-ordination challenge; producers, this is a way to add some flavour and spice to the standard Shuffle Groove.

A cool variation of this is to take the hi-hat notes and move them to the ride cymbal whilst keeping the hi-hat pedal in the same place. At slower tempos, this creates a laid-back feel that can really help complement a song and bring it to life.

Groove creation: Challenge two

Your challenge

Your challenge to help you get the most out of this chapter is to select a pattern from the list shown and then use numbers from our 8th note triplet grid. From combining these different Shuffle/triplet patterns and bass drum notes, you will be able to create a whole host of Shuffle/triplet-based grooves.

Just as at the end of Chapter 1 Starting Essentials, we were presented with different hi-hat options that can fit onto our grooves. In this chapter we have now been exploring swung rhythms that are triplet-based. These are the different hi-hat (or ride) options that we have come across so far in this chapter. Each of these can be mixed and matched in the context of the grooves found in this chapter. This gives us a variety of new Shuffle-based grooves and variations of feels we can use in our music creation.

Option 1 – The Shuffle

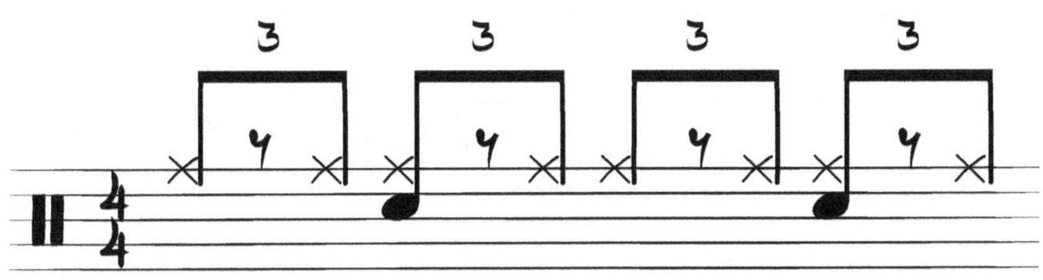

1 & a	2 & a	3 & a	4 & a

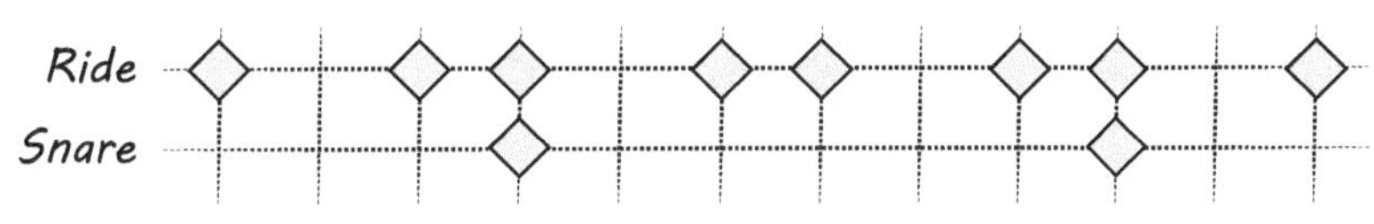

Option 2 – The Texas Shuffle Rhythm

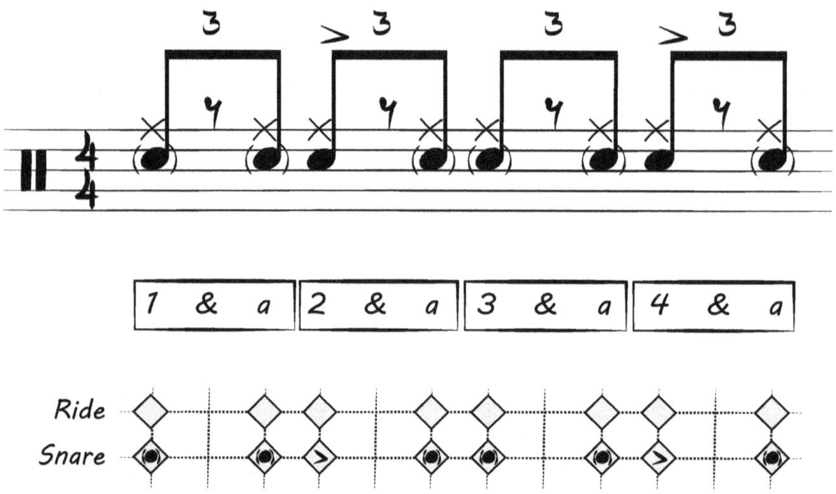

Option 3 – The Rock Shuffle

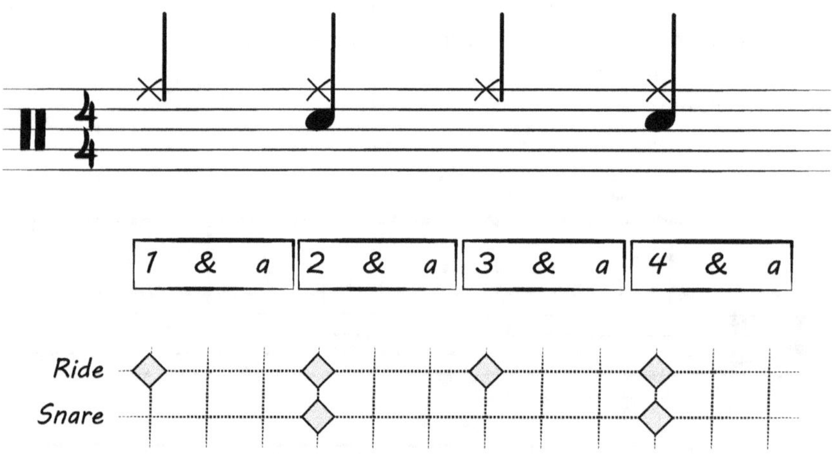

Option 4 – Jazz Shuffle

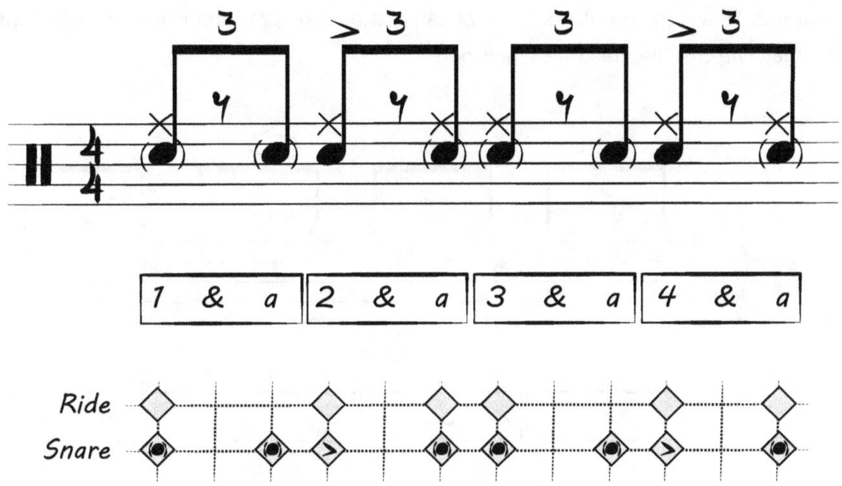

Option 5 – The Stepped Shuffle

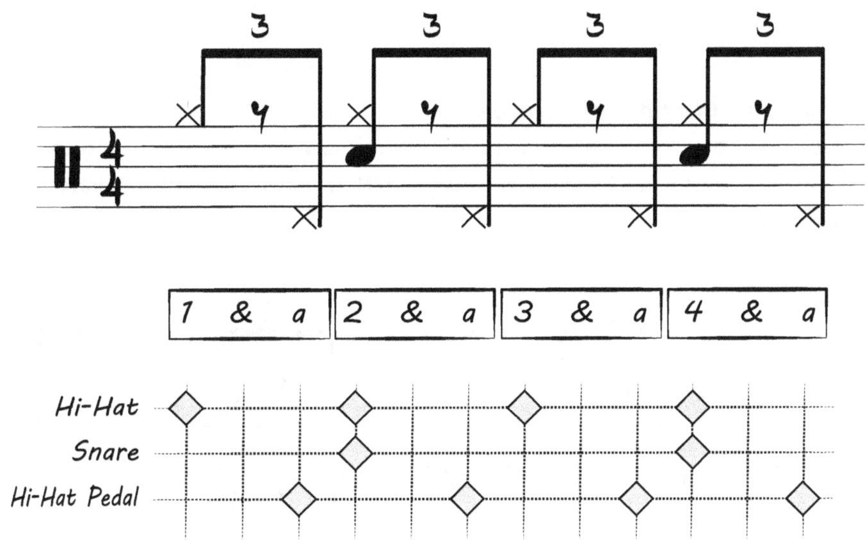

NB Please note that all of these can be play on hi-hat and ride (or even crash cymbal or floor tom). Experiment and find what works for you.

Drum Groove Creation

8th note triplet grid

In a similar way to how we used our 16th note grid in Chapter 1, we have a triplet-based one for Chapter 2. This is the numbers 1–12, because there are 12 triplet notes in a bar. Note that Option One, The Shuffle, has been included.

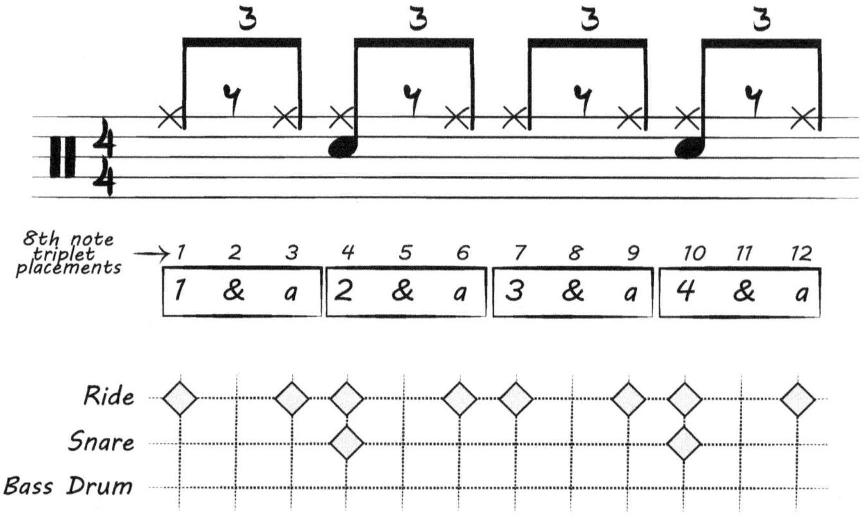

Groove creation: Challenge two example

Now, let's take some of these numbers to create a new groove whilst using the Shuffle pattern on the hi-hats and a backbeat on the snare. Here I have selected, 1,5 and 9 from our 8th note triplet grid

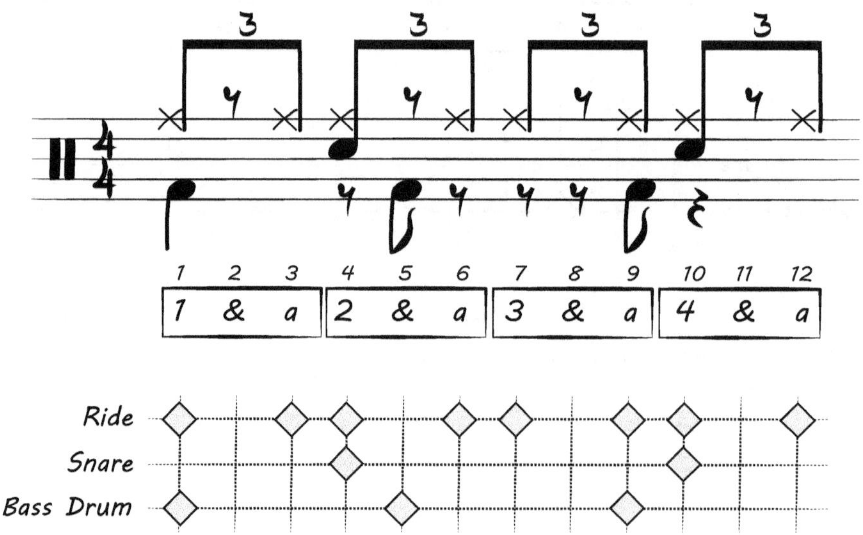

Your challenge

This is just an example. For your groove creation exercise, take these different ride/hi-hat and snare patterns from this chapter and select bass drum placements from our 8th note triple grid to create new grooves.

Extra content: Useable triplet fills

I have sprinkled some extra content throughout this book, where I briefly discuss fills that could work alongside these grooves. It is worth noting that I could (and at some point possibly will) write a whole book on Drum Fills and Drum Fill Construction. But this book is for Drum Groove Creation, so the majority of the book will focus on that subject. For now, here are some useable fills that I have found work in a variety of musical scenarios.

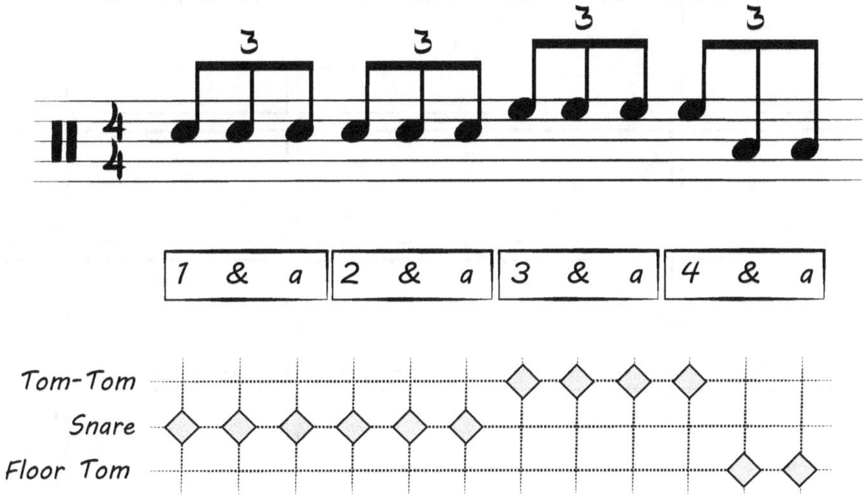

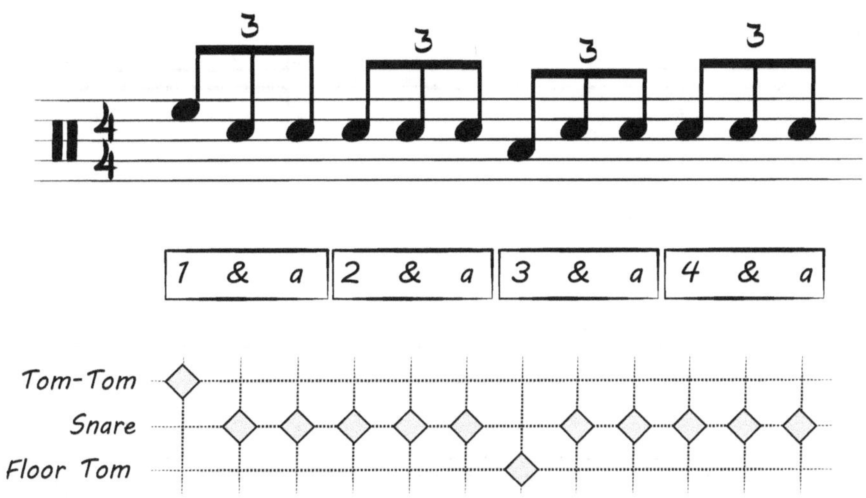

Drum Groove Creation

When	What	Tempo	Notes/Reflections

Chapter 3 R&B and Early Rock 'n' Roll

Genre overview

So far, we have discussed our Starting Essentials, the 'Money Beat', ghost notes, bass drum placements and some of our different shuffle variations and ways to construct them. We are yet to really dig into Early Rock 'n' Roll and Rhythm and Blues (R&B) music. Some may see this as a movement of the past and not now worth bothering with, but I disagree. Many of these artists, drummers and their grooves have inspired the music that we listen to today. By extension, one may surmise that because it has impacted the music we listen to today, it therefore has an impact (subconscious or not) on the music we create today. In addition, one of the hardest things about capturing this genre is the 'feel'. This is even more challenging if you are programming these grooves.

Most of us today have grown up listening to music that is played 'straight', and this reflects in our drumming and music creation; it is almost hardwired into our drumming and musical DNA (not an officially recognised medical term in case you were curious).

However, back in the early 20th century, Jazz was all the rage and notable Early Rock 'n' Roll /R&B drummers such as Earl Palmer and Charles Connor would have grown up listening to swung music. As time progressed, however, the music gradually began to 'straighten out' and the backbeat was pioneered into popular music. The result of music getting gradually straightened out and the DNA of swung music in these drummer's veins resulted in a feel between straight and swung.

Key artists to listen to

Elvis Presley
Little Richard
Fats Domino
Bo Diddley
Jerry Lee Lewis
Ray Charles

DOI:10.4324/9781003585169-3

Key drummers

Earl Palmer
Charles Connor
Ronnie Tutt
Dominic Joseph (DJ) Fontana
Hal Blaine

It is worth noting that some of some drummers have played on early R&B, Gospel and Blues records. To me, there is no clear-cut divide between these genres; instead it's a big melting pot of influential drummers who helped shape the artists and the music around them. So, before anyone gets too tied up in knots over dividing artists and drummers according to singular genres or styles, let's just enjoy their playing and their contribution.

Groove 17: The Bo Diddley beat

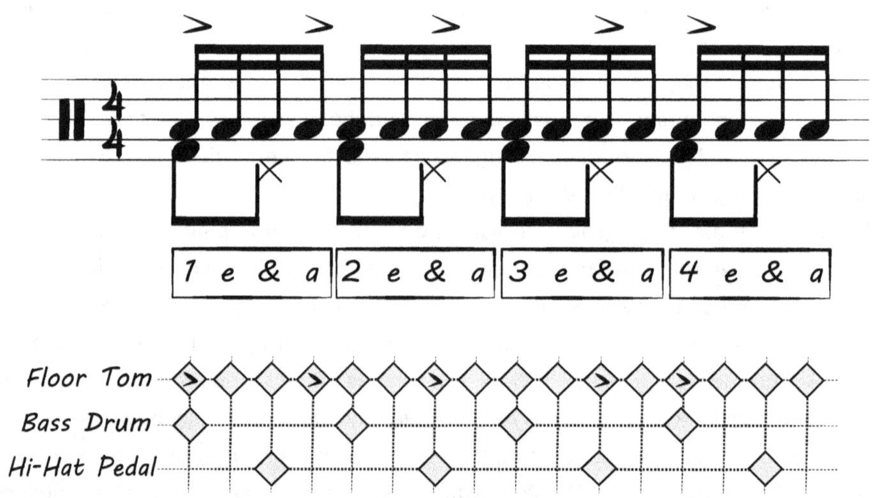

I felt like this one could have quite easily slotted into the Blues chapter; however, I also thought it was quite a fun groove to open the Early Rock 'n' Roll/R&B chapter with. It is the first groove we are encountering that moves away from the ride cymbal or hi-hat, with both hands playing on the floor tom. The key thing here is the accents; they form a 3:2 clave rhythm. Try playing/listening along to the song *Bo Diddley*, by (you guessed it) Bo Diddley.

The importance of Earl Palmer

Earl Palmer is perhaps not one of the most familiar names these days amongst drummers and artists, but he made an undeniably significant contribution to music as we know. Often credited with pioneering the backbeat, Palmer's playing on Fats Domino's *The Fat Man*, amongst other tracks, helped define what came to be seen as the norm. From Earl Palmer and the spread of this style of music, the backbeat gradually made its way into most records of the era.

When asked about his part on *The Fat Man*, Palmer said, *"That song required a strong afterbeat throughout the whole piece. With Dixieland you had a strong afterbeat only after you got to the shout last chorus. ... It was sort of a new approach to rhythm music."*[1]

Earl's playing was influential, and he became one of the most in-demand session musicians of the 20th century. His style and flavour of drumming can be heard on records by Frank Sinatra, Little Richard, Fats Domino and Professor Longhair, among many others. I would highly recommend listening to his catalogue of recordings. This is not to emphasise or showcase how 'flashy' his drum parts were, but more to highlight how beautifully they fit and complement the songs.

Groove 18: Snare groove with Backbeat

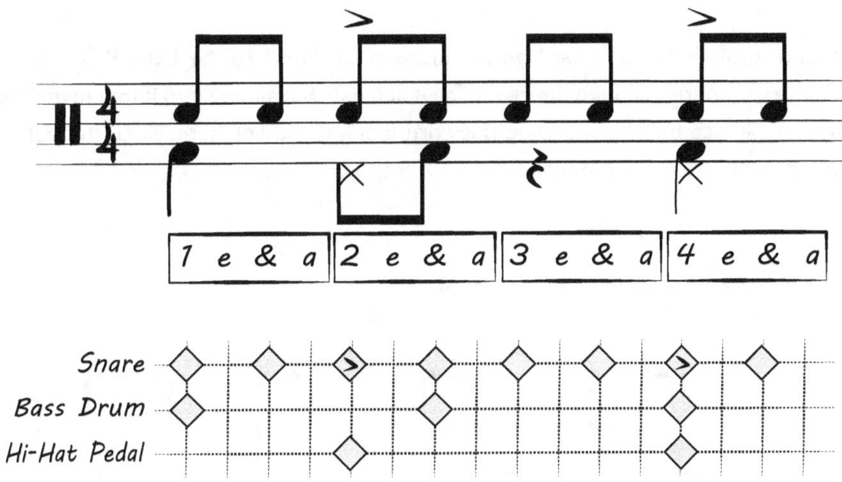

In a sense you could consider this as the 'Money Beat', but played on the snare drum. This same groove can be found in genres as diverse as Glam Rock and Country Music. It is sometimes referred to as a 'train beat'. We first see it crop up and played by Earl Palmer in early Rock 'n' Roll. To hear this style of groove in action listen to *I'm Walkin'* by Fats Domino and also *Slippin' & Slidin'* by Little Richard. Variations of this beat (which utilise different accent patterns) can also be found in some of Palmer's other works, such as *King Kong* by Big T Tyler.

In this example I have added a bass drum pattern commonly used by Palmer. It can be found elsewhere in music can is sometimes called the Baiao, the Mambo bass drum pattern or the three side of the Clave.

1 https://en.wikipedia.org/wiki/Earl_Palmer

Groove 19: Backbeat on the cross-stick

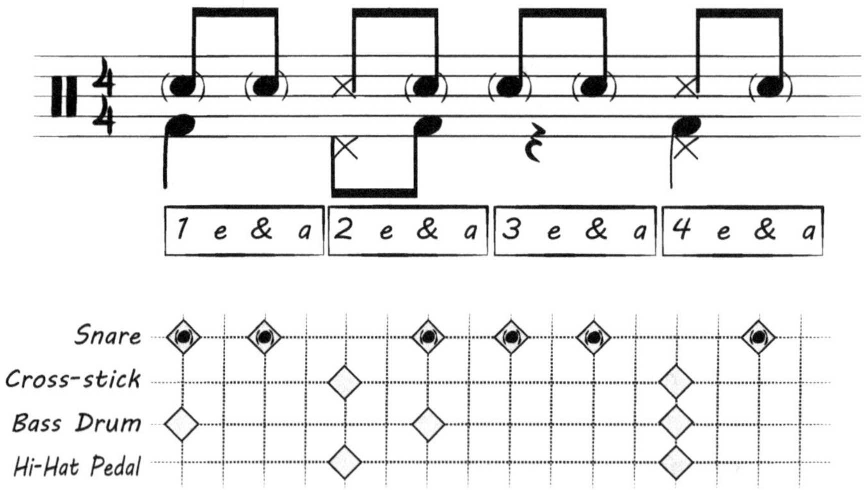

This style of groove can be heard on the first part of *Rip It Up* by Little Richard. This is another one to experiment with the swing, as it still has that slight lilt to it that gives it some extra flavour. A nice boomy-sounding bass drum can work well here. Keep the volume of the snare drum hits low, so it doesn't overpower the cross-stick.

Groove 20: Unison between snare & hi-hats

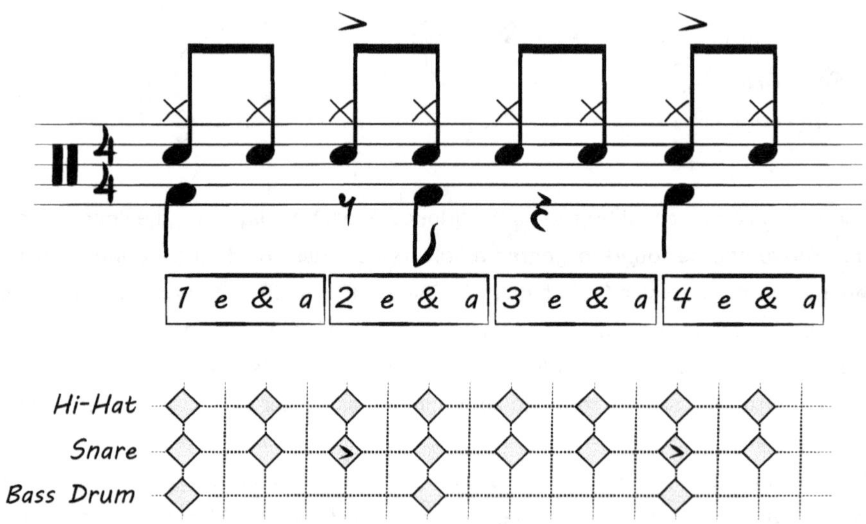

It's odd in a way how we began with Starting Essentials, but that, as we dig further back to understand our groove creation journey (and how to create new things), we see just so many similarities between grooves being played today and others from the 1950s.

This groove utilises a unison between the snare and the hi-hats; they are, in essence, both playing 8th notes with an accent on two and four to give us our backbeat. These influences later sparked the likes of the playing of Led Zeppelin's drummer John Bonham on the band's track *Rock and Roll*.

Groove creation: Challenge three

Groove creation: Challenge three has two parts

1. Take the Bo Diddley beat. Orchestrate it on different parts of the drum set and experiment with the amount of swing on it. Move the accents to different drums and see what you can come up with.

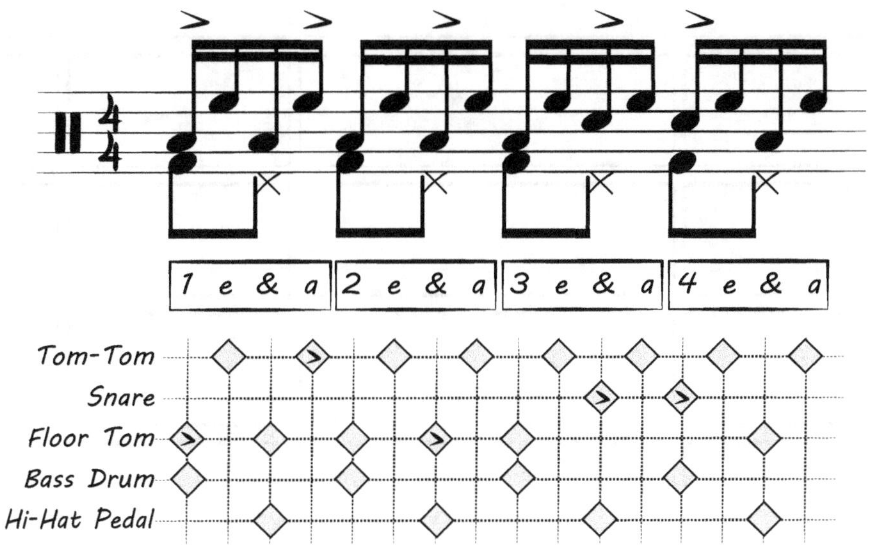

2. In a similar way to Chapter 1, we took the 'Money Beat' and added different bass drum notes across our 16 placements. Here we will do the same but for Groove twenty.

When	What	Tempo	Notes/Reflections

Chapter 4 Soul

Genre overview

Soul overview

Emerging from Black communities in the United States, Soul music blends elements of Rhythm and Blues and Gospel to create a unique sound. A large element of Soul music is in the lyrical content; like many other genres, there are a plethora of songs about heartache and romance.

In addition to this, Soul artists such as Marvin Gaye and Aretha Franklin used their music as a voice during the Civil Rights Movement in the 1960s. Their powerful and meaningful lyrics are supported by a driving rhythm section that allows the audience to move and dance together as one. Original 1960s Soul music has influenced other genres, such as Funk, Disco and Neo-Soul.

Key Soul drummers

Uriel Jones
Benny Benjamin
Richard 'Pistol' Allen
Al Jackson Jr.
James Gadson
Bernard Purdie
Howard Grimes

Motown overview

Motown in and of itself may be considered a distinctive style of Soul music. Another pillar of Black communities, Motown gets its name from the Motown Records label which was based in Detroit (also widely known as the Motor city). Its memorable melodies and catchy hooks captured the attention of millions, and the distinctive bass lines and signature drum sound helped give the genre a unique and memorable sound.

DOI:10.4324/9781003585169-4

Some key Motown artists

Diana Ross and the Supremes
The Temptations
Stevie Wonder
Smokey Robinson and the Miracles
Martha Reeves and the Vandellas
The Four Tops
Marvin Gaye

Stax Records overview

Stax Records was known as a mixed-race record label. It notably had a mixed house band of Booker T and the MGs, and it also promoted both Black and white musicians. This was one of the elements that contributed to the distinctive Southern Soul Memphis sound. These records have a blend of Black Gospel and early Rhythm and Blues, as well as a slightly grittier sound than that often found on Motown records.

Some key Stax artists

Otis Redding
Sam and Dave
Booker T and The M.G.s
Isaac Hayes
Steve Cropper
Rufus and Carla Thomas

Groove 21: The classic Motown groove

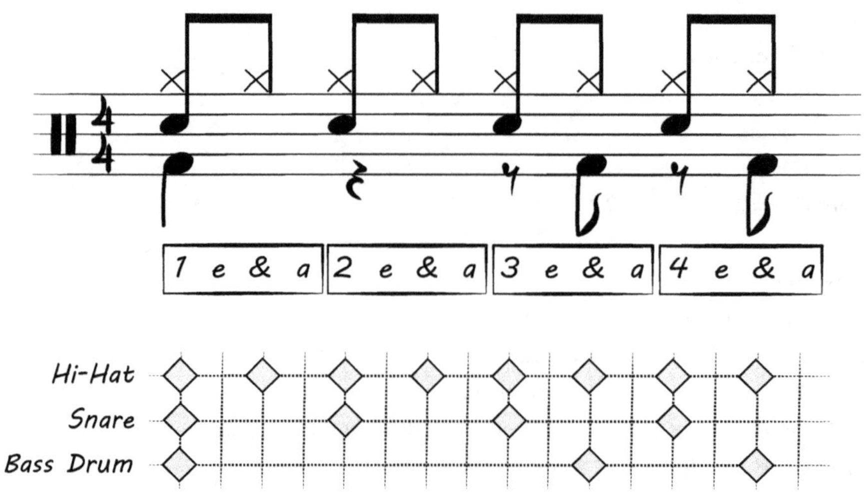

When someone refers to 'THE Motown beat', this is what they mean; this is the signature Motown groove! This groove is the foundation of that Motown sound. Quarter notes on the snare are what you want to 'pop'; be careful not to be too heavy on the hi-hats. It's very easy to turn this into a heavy rock beat.

Groove 22: Alternative Motown groove

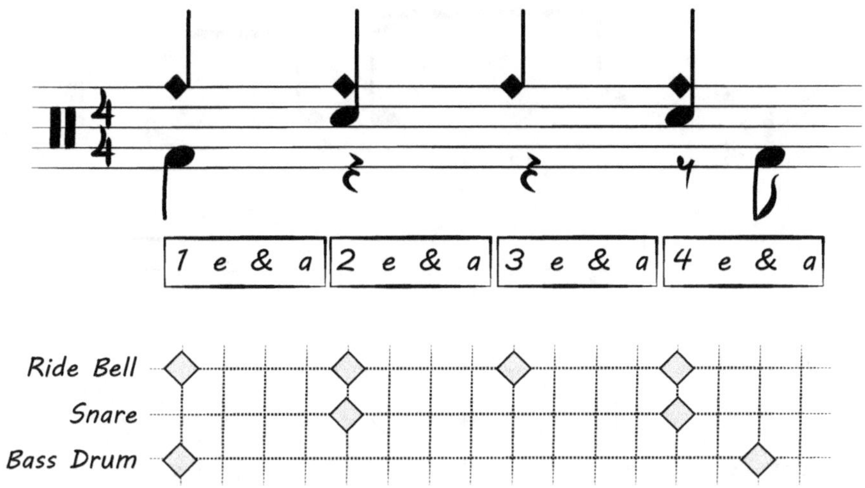

This is a nice alternative to the classic Motown groove that we have just encountered. This groove can be heard, for instance, on the introduction of *Ain't Too Proud to Beg* by The Temptations.

Extra content: Useable Motown fills

Well, if we have a signature groove, then we need signature fills. Here are a few variations on Motown fills. Some of these may lend themselves to different tempos, i.e., playing the six stroke roll at 140bpm is quite a challenging feat. As a result, some of the 16th note-based fills may be better placed in that kind of music scenario when the track is at a slower tempo.

Motown fill example one

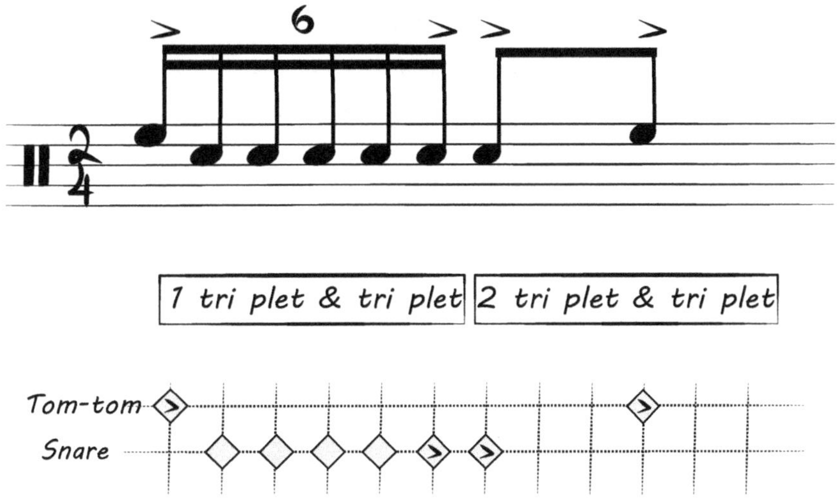

Motown fill example two

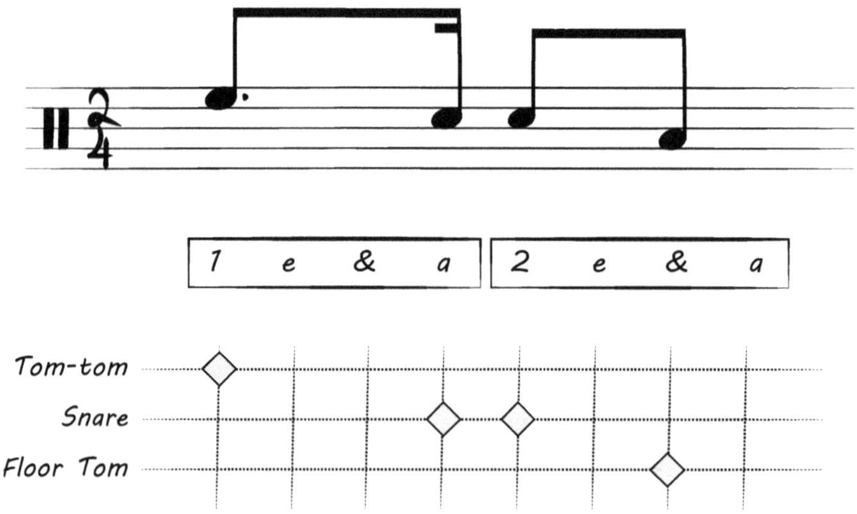

Groove 23: Southern Soul example one

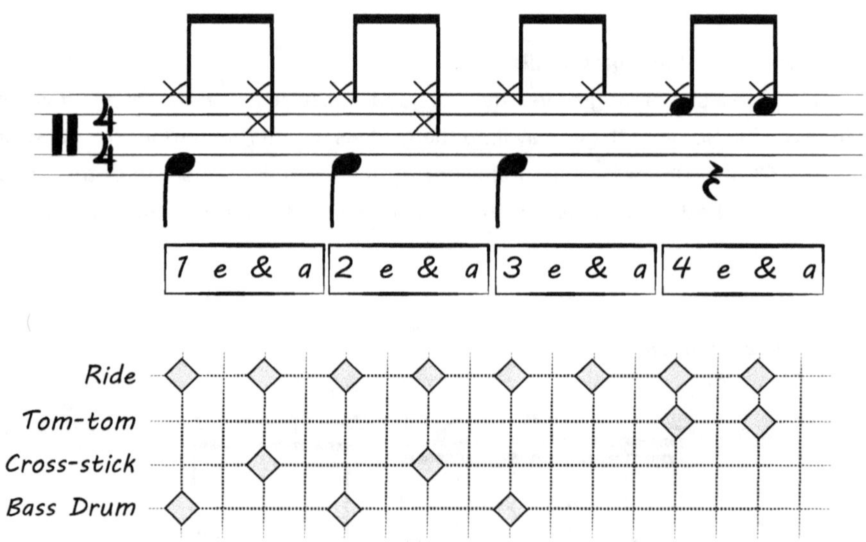

This is the first groove that have encountered which is played in the style of Al Jackson Jr. In my opinion, Jackson was a very underrated drummer who has played on countless hit records and really supported the artists whom he was playing for. I like the feel to this one and the two tom tom hits at the end give it a slightly Latin feel.

Al Jackson Jr is a great example of playing for the song and for the music. Just because you can play/programme a complex drum part, this doesn't mean it always works with the song. Always listen out for what the other instruments need and how the drums can support them.

Groove 24: Southern Soul groove example two

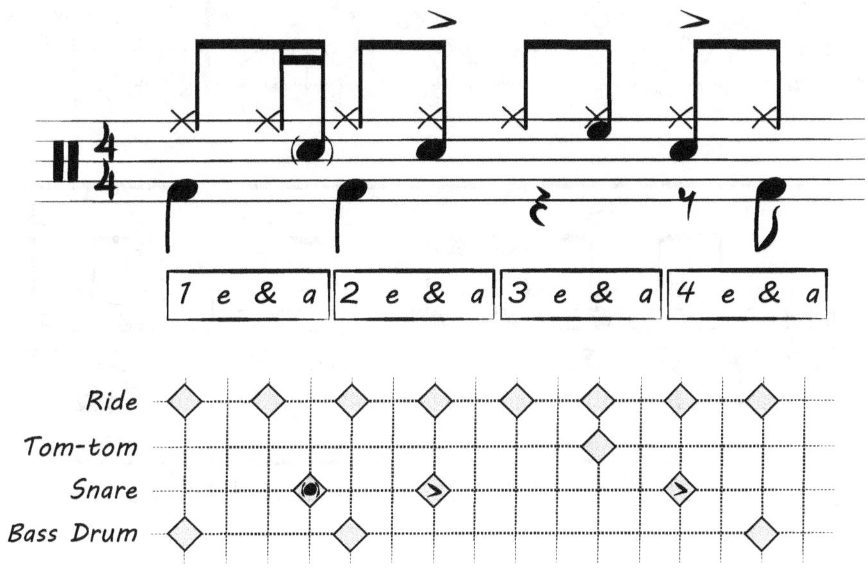

I like this other Al Jackson Jr groove as it has a nice interplay between the snare and tom tom, which creates a unique-sounding melody. This gives us an example of how we can begin to incorporate some other elements of the drum set into our grooves.

Groove 25: The Amen Break

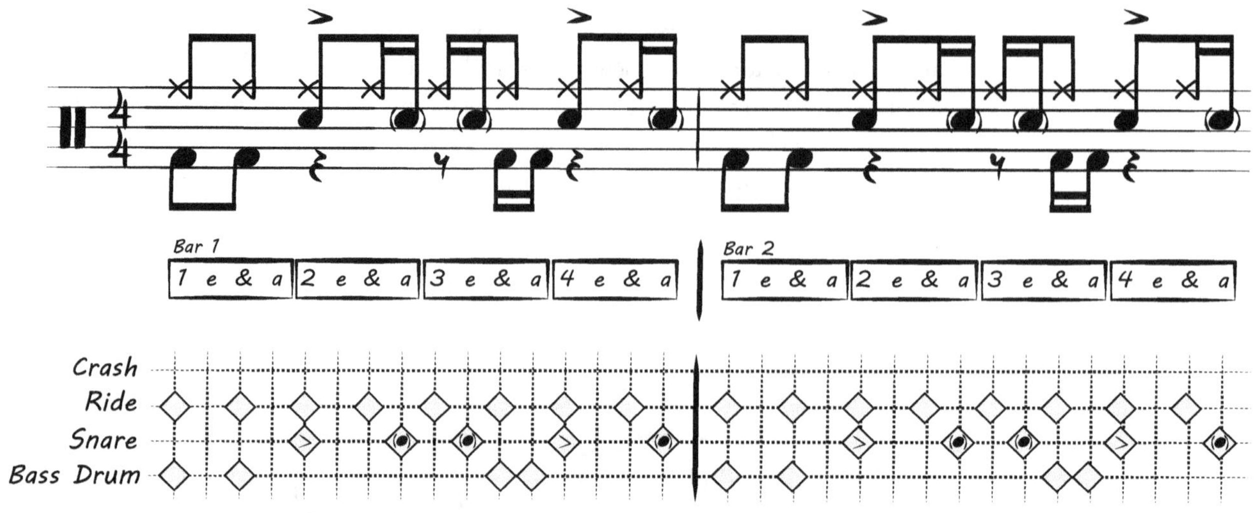

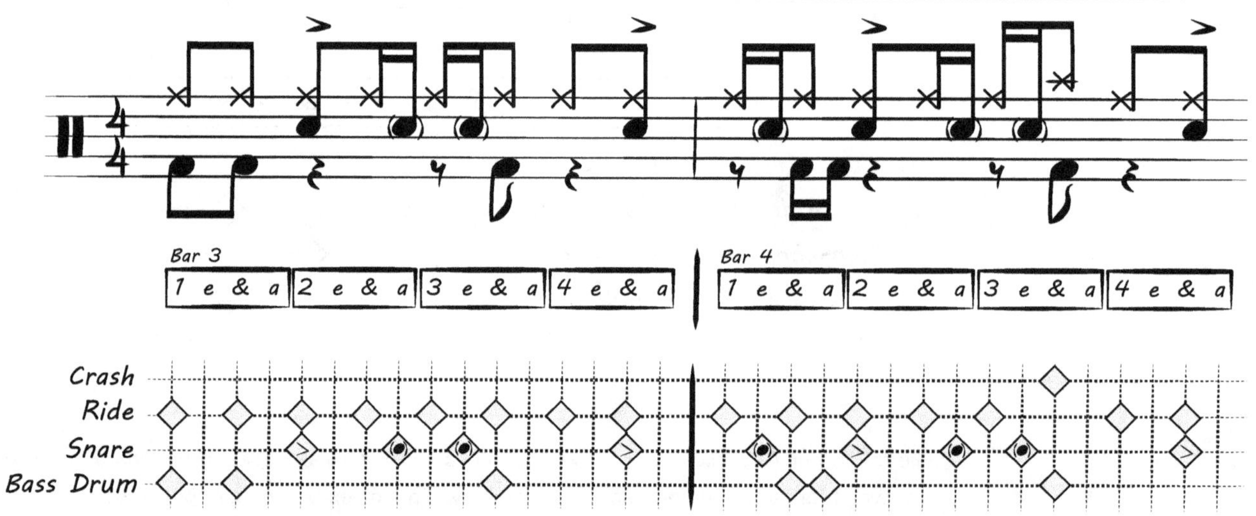

One of the most (if not *the* most) sampled drum breaks of all time. This is from the song *Amen, Brother* by The Winstons. The drummer who played on this recording was Gregory Coleman. Currently, according to whosampled.com, *Amen, Brother* has been sampled on a whopping 6768 songs. As well as listening to the original song and drum break, I recommend checking out this website and listening to some of the tracks it can be heard on.

For producers who want to know how to create breakbeats and where they originated from... look no further than this one. Another great reference point is the *Funky Drummer* by James Brown (discussed in Chapter 5).

Groove creation: Challenge four

Expanding on these rhythms

In this chapter we have introduced some important and key grooves. But now we want to manipulate these further and put your own spin on them. So, in this challenge we are going to concentrate on altering the feel of the hi-hats and ride.

Here I am reintroducing the hi-hat options from Chapter 1. Refer back to the images Chapter 1 hi-hat options if you need a reminder.

Option 1 – Straight 8th Notes
Option 2 – Digging In
Option 3 – Upbeat Feel

Your challenge

Take the grooves in this chapter and apply these different hi-hat/ride patterns to see how it alters the feel and the difference it makes to your grooves.

Pushing yourself further: to get even more out of this chapter and concept, try putting these rhythms on the ride cymbal but with the accents being played on the bell of the cymbal.

In addition, I recommend experimenting with replacing the normal backbeat on the snare drum with a cross-stick. Again, giving us more options and more variety. Learning and growing all of these tools will help in our ability to create our own grooves.

When	What	Tempo	Notes/Reflections

Chapter 5 Funk Music

Genre overview

So, this is where things start to get a little bit spicy, with the introduction of some very interesting and creative grooves. But first, a little bit of history into this genre and some of the key and notable artists.

James Brown

Known as 'The Godfather of Soul' and 'The Hardest Working Man in Show Business', James Brown is one of the pioneers of the Funk genre. Influenced by R&B, he helped bring a brand-new identity to this genre of music. His music often emphasised 'the one' (beat one in the beat) for effect and giving his music an unmistakable dance quality. This is coupled by hypnotic rhythmic guitar and a fabulous horn section.

James' most notable drummers were Melvin Parker, Clyde Stubblefield, Jabo Starks and also, on occasion, Brown himself (check out the song *Limbo Jimbo*).

The Meters

Another Funk band (with a very different vibe and feel) is The Meters, with Joseph 'Zigaboo' Modeliste on drums. For an example of how the previously discussed 'New Orleans second line feel' can be used in a Funk context, listen to The Meters back catalogue.

Tower of Power

This band has been through a number of changes in line-up over the years, but their most heavily featured drummer has been David Garibaldi. This drumming style works seamlessly with the bass guitar and is often made up of unique linear style grooves. For more information on linear drumming, explore Chapter 12 on 'Creating linear drumming grooves'.

Other notable Funk artists

George Clinton's Parliament/Funkadelic
Bootsy Collins
Sly & The Family Stone
The Isley Brothers

DOI: 10.4324/9781003585169-5

Key drummers

Claton Fillayu
Melvin Parker
Clyde Stubblefield
John 'Jabo' Starks
Joseph 'Zigaboo' Modeliste
David Garibaldi
Greg Errico

Groove 26: The displaced backbeat

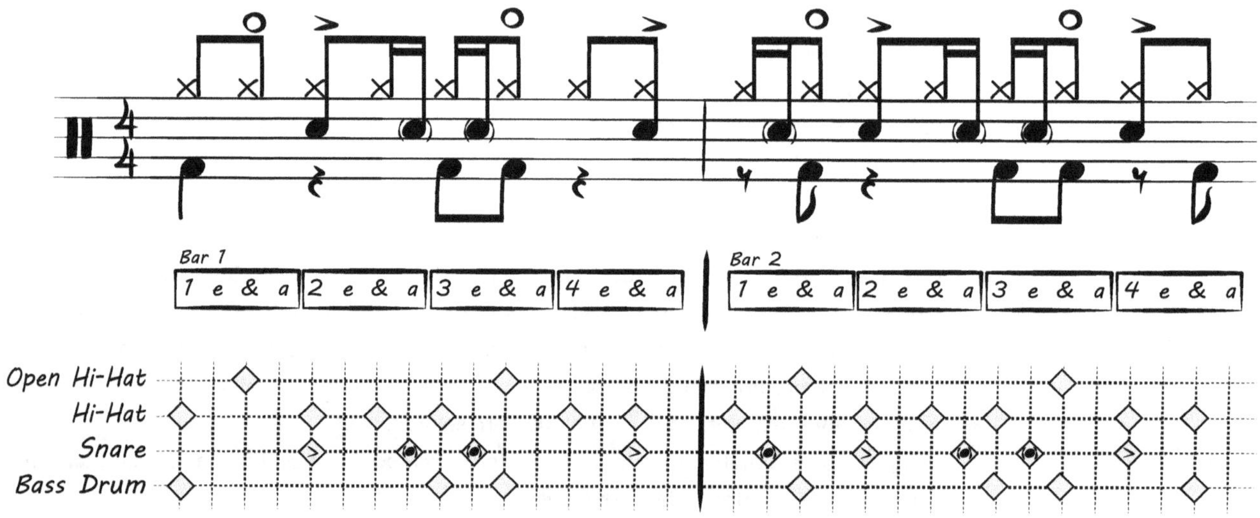

In Clyde Stubblefield's 1967 classic groove, which features on James Brown's song *Cold Sweat*, note the inclusion of open hi-hats with the left foot on 'the and of 1' and 'the and of 3'. Another Funky tool we are introduced to here is the displaced backbeat. A displaced backbeat essentially means that the backbeat is not falling on both beats two and four in the bar. Here we see it moved to the 'and' of beat four.

Groove 27: Quarter note Funk

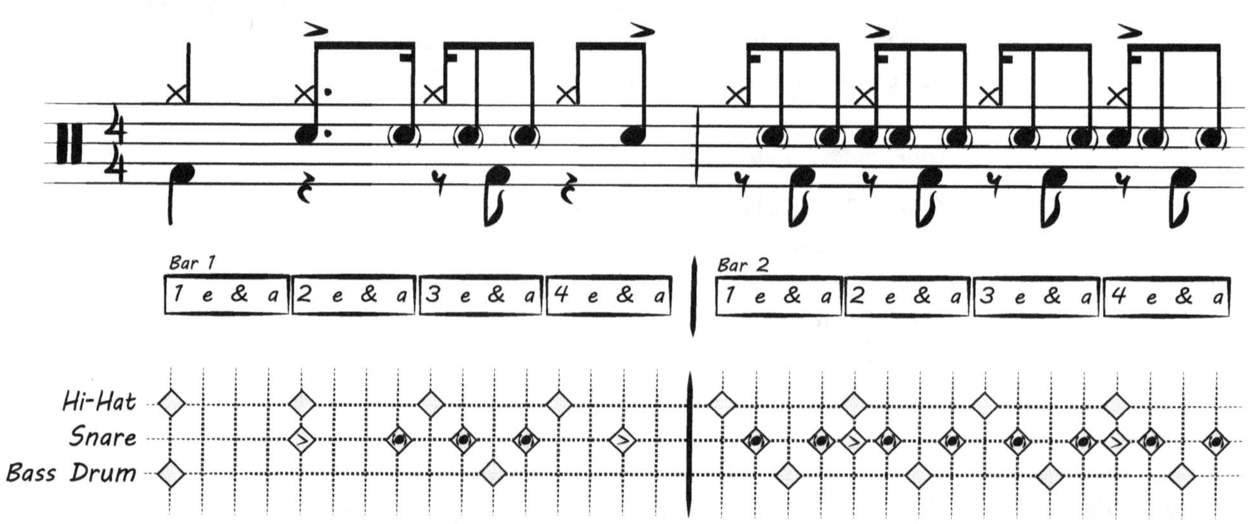

On the list of my top five James Brown grooves, *Mother Popcorn* holds a key spot.

Note here that, in contrast to the previous example, the hi-hats are playing quarter notes. The snare drum is quite busy here, with a lot of ghost notes in the second bar.

Drummers: In the second bar, I would recommend coming down for the backbeat in your right hand and leaving out the hi-hat note. If you are a left-handed player, you could reverse this.

Groove 28: Skipped hi-hats

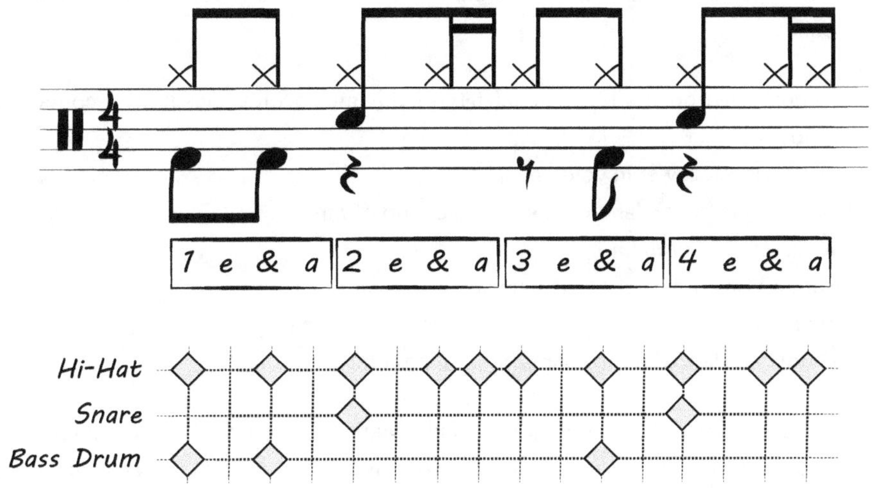

Here we encounter a Jabo Starks-type trait that he would often play on the drum set. Essentially, after the backbeat we have two swung hi-hat notes. This is actually similar to the Jazz ride cymbal pattern that we have previously encountered in this book. I would recommend swinging these notes slightly, so it doesn't sound too robotic.

Groove 29: *Funky Drummer*

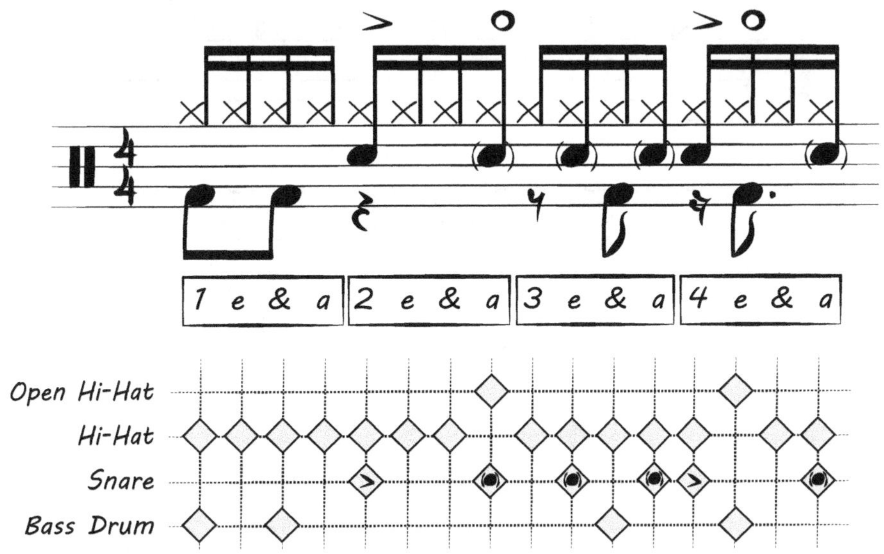

Drum Groove Creation

I referenced this earlier in the book, when discussing the 'Amen Break'. *Funky Drummer,* released in 1970, contains one of the most iconic drum breaks ever heard. This has been sampled and replicated on so many records over the course of the last fifty or so years. It is a stone-cold Clyde Stubblefield classic.

Groove creation: Challenge five: How to make drums Funky

I wanted to compile a short bullet pointed list that brings together some of the key characteristics of the Funk grooves we have seen above. Note that you don't have to use all of these in a single groove; try some different combinations and always think of what fits the music best.

- A repetitive groove with minimal fills in the music (this makes it become almost hypnotic)
- Hi-hats slightly lower in volume
- The open hi-hats on 'and of one' and the 'and of three'
- The displaced backbeat
- Skipped hi-hat notes after the backbeat
- Ghost notes
- 16th note Hi-Hats with the snare drum on beat three.

As previously discussed, different hi-hat options can change and alter our feel for different effects. Here are three hi-hat options that are commonly found within Funk music. They can, however, be applied to any musical situations.

Funk hi-hats option A – Opening the hi-hats

This particular hi-hat pattern has been used by Melvin Parker, Clyde Stubblefield and Jabo Starks.

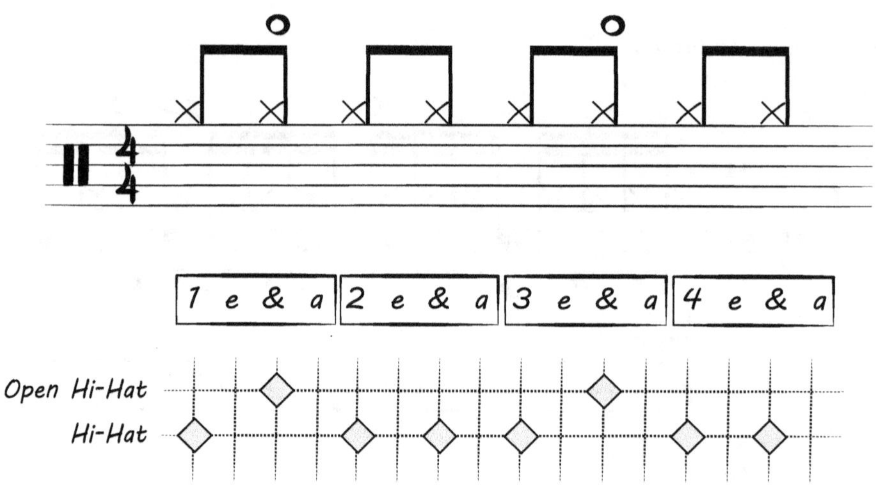

Funk hi-hat option B – Skipped notes

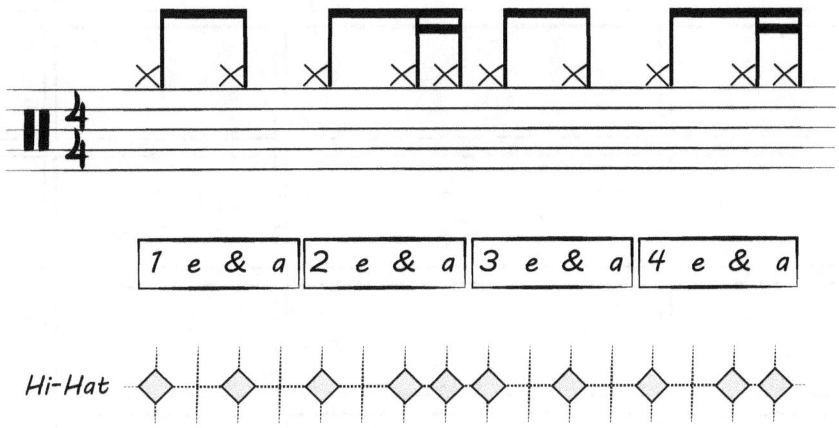

Funk hi-hat option C – Skipped notes with openings

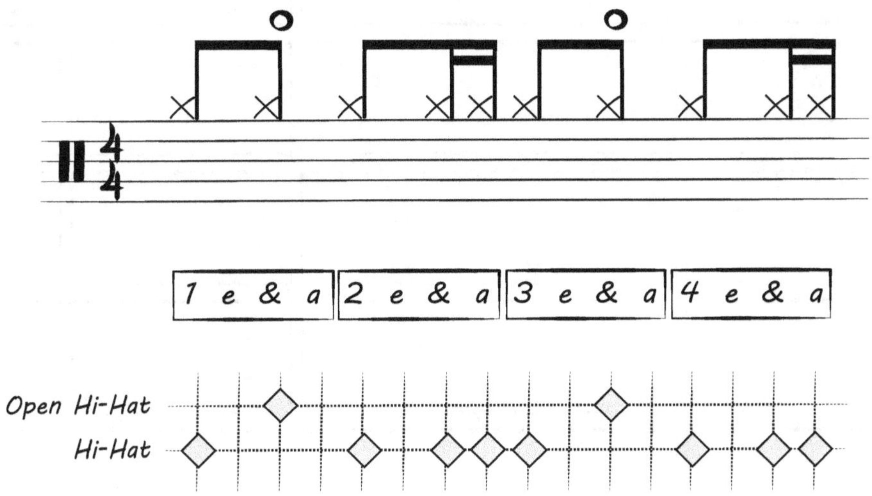

Using the characteristics outlined above, it is your job to start to make some of your own Funk drum grooves.

When	What	Tempo	Notes/Reflections

Chapter 6 Hip-Hop

Genre introduction

Hip-Hop is a subculture that involves not only music (although notably for MCing/rapping), but also scratching turntables, break dancing and street art. This genre is known for its almost hypnotic and strong drum grooves. Hip-Hop emerged in the early 1970s in New York City; since then, it has spread across America and the world. This music often features sampled Funk tracks and drum grooves. It has also birthed other subgenres such as Grime, Gangsta rap, Drill and Trap.

Key artists

As mentioned above, this genre has birthed many subgenres to explore, but here are a few Hip-Hop artists to begin your exploration:

Tupac Shakur
The Notorious B.I.G
The Roots
Jay Z
Nicki Minaj
NWA
Queen Latifah

Key drummers

So, this is an interesting one. In previous chapters, I have discussed key and influential drummers to the set genres. In this genre, however, many of these artists use programmed drums or early drum machines. It is also worth mentioning again how many old Funk and Soul records have been sampled to help create the feel of this new genre.

DOI:10.4324/9781003585169-6

Groove 30: Hip-Hop example one

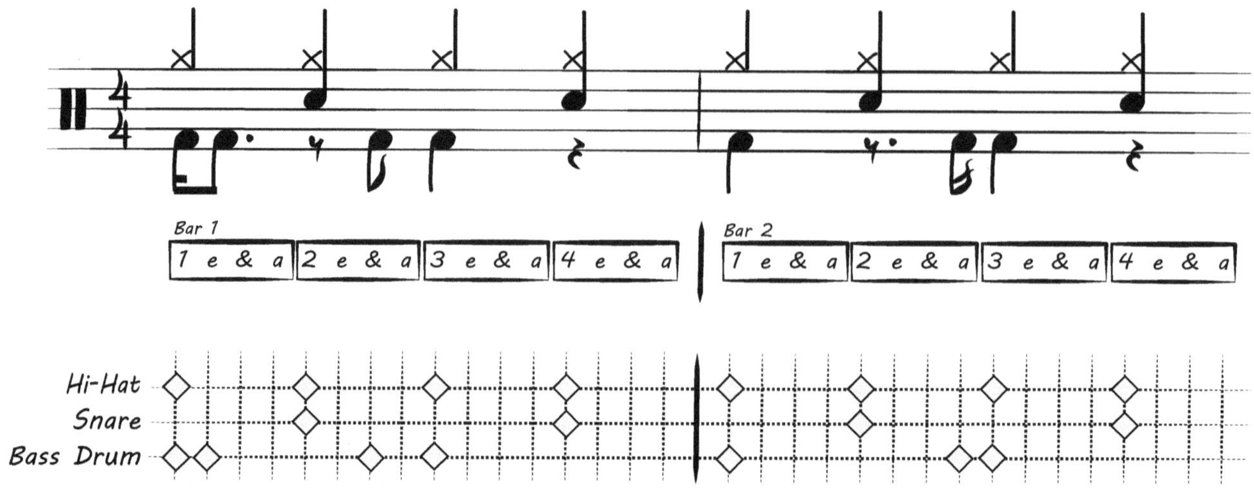

Here is something that we have explored only briefly in our drum grooves; the hi-hats playing on the quarter note pulse. This one derives from Grandmaster Flash & The Furious Five's song *The Message*. The recording features some additional hi-hats throughout the piece. This example captures the essence and feel of the groove.

Groove 31: Hip-Hop example two

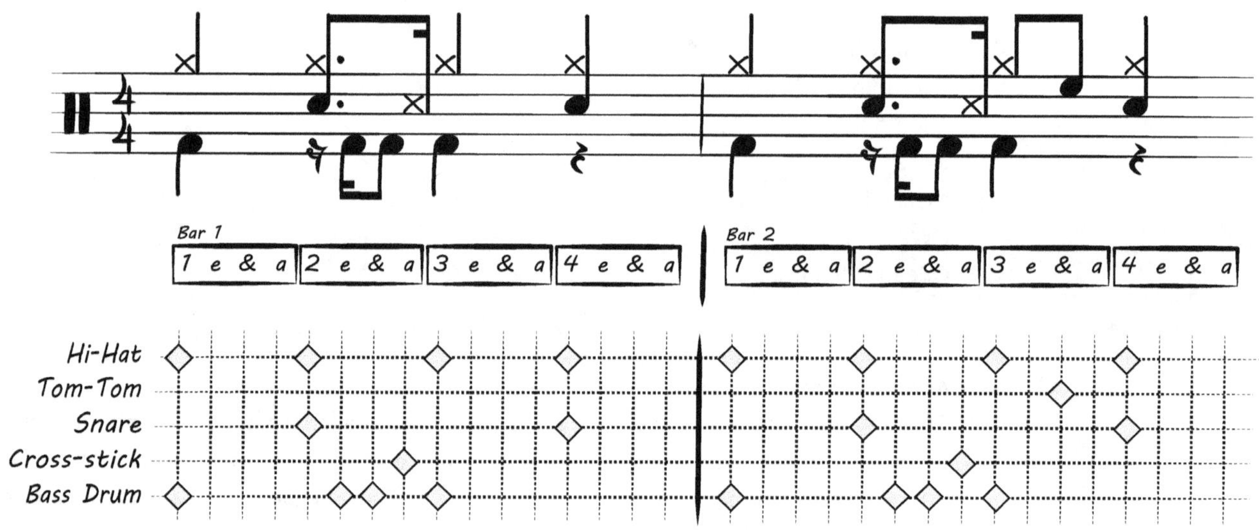

This is based off mid-1990s Hip-Hop/Gangsta Rap. It is the kind of groove that was often programmed on a drum machine. For listening check out *Juicy* by The Notorious B.I.G.

Note the quick change between the snare and the cross-stick as well as the additional tom tom hit in the second bar.

Groove 32: Hip-Hop example three

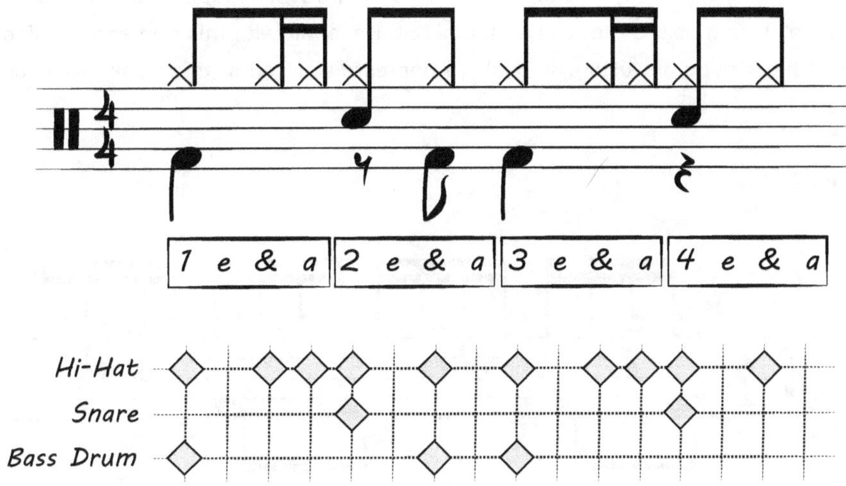

This example features a new hi-hat pattern. This is a relatively straightforward groove, but when repeated it becomes hypnotic. A lot of Hip-Hop music is about the lyrics and the rap. By maintaining a consistent unwavering drum groove, we support this and allow the lyrics to shine through.

Groove 33: Trap

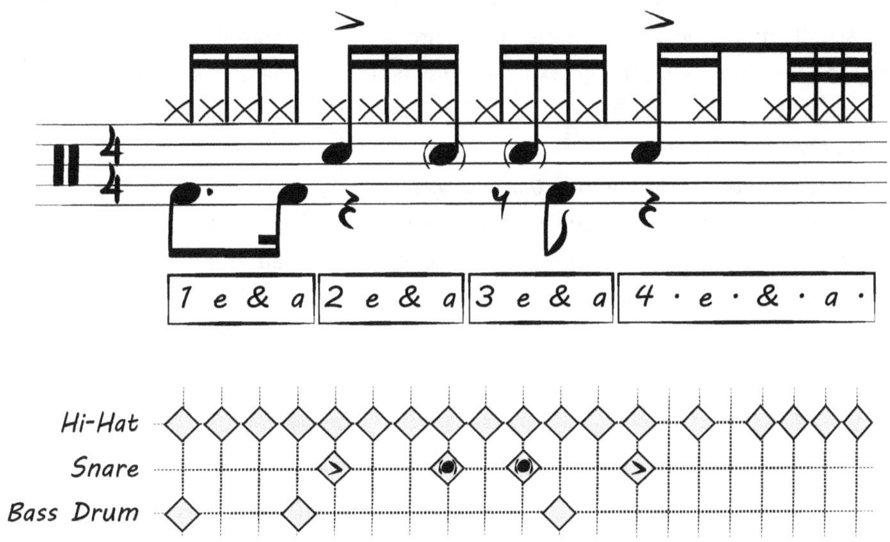

Drum Groove Creation

Trap is one of the subgenres of Hip-Hop that I wanted to showcase within this chapter. It is often arranged at slower tempos to allow a laid-back groove. In addition to this, a key characteristic of Trap music is the rolls on the hi-hat. Have fun with this one and consider the sound of the kit you are using. As with all genres discussed in this book, go listen and explore.

Groove 34: Grime

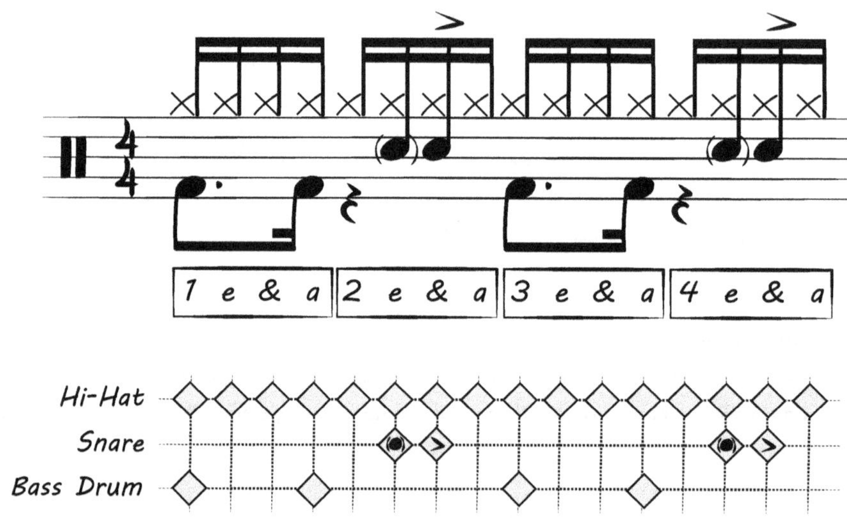

Another subgenre of Hip-Hop, Grime has emerged over the last decade or so; through artists like Stormzy, it has hit the mainstream. Sometimes Grime tracks have a tendency to have slightly more complex drum parts that alter and vary more than old school Hip-Hop. Listen to Stormzy's back catalogue to gain an overarching awareness of this.

Groove creation: Challenge six

Cross-stick and backbeat combination grooves

Here we are combining two of the elements found in this chapter. We have the cross-stick and backbeat from Hip-Hop example two and the hi-hat pattern from Hip-Hop example three. If we combine these together, we have a brand-new Hip-Hop-style ostinato.

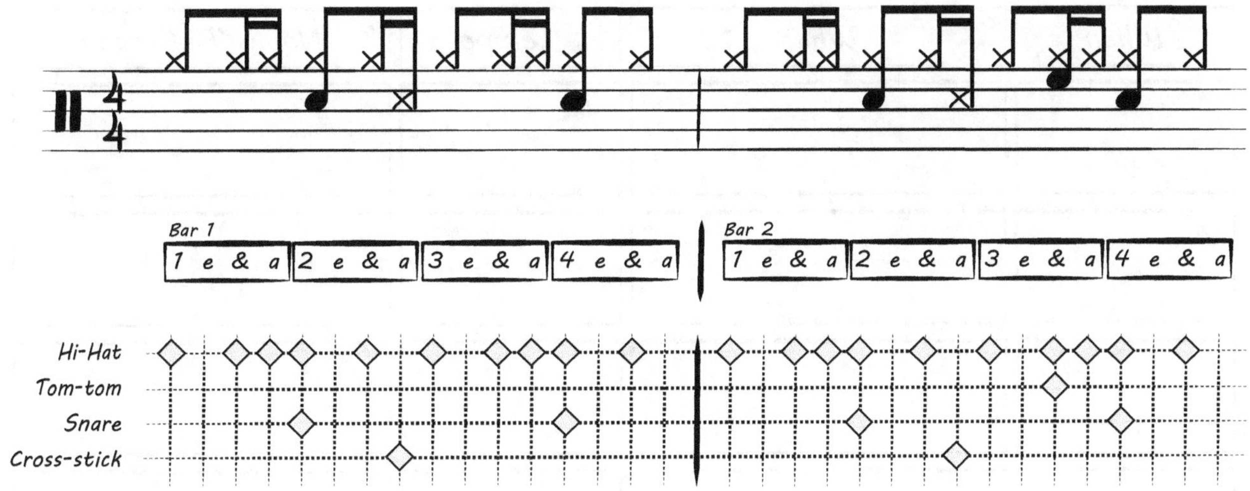

To now add our bass drum to the mix, let's start with a simple Disco/House-inspired four to the floor. This is your starting point. Next start adding and constructing your own bass drum patterns. *Remember our 16th note grid from Chapter1.*

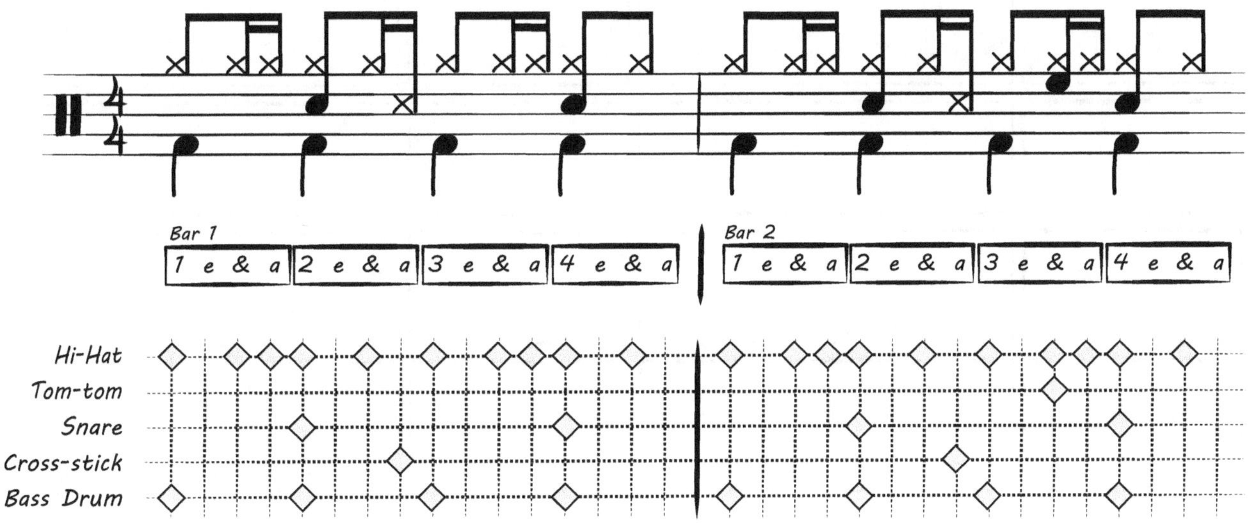

When	What	Tempo	Notes/Reflections

Chapter 7 Jazz

Genre overview

The early roots of Jazz began in the great city of New Orleans in the late 19th and early 20th centuries and continues to evolve. Often driven by a heavy swing rhythm and large elements of improvisation, it has birthed many subgenres and is universally recognised as an incredibly important aspect of the heritage of western music. This later gave way to the Big Band era, with famous artists such as Frank Sinatra, Dean Martin and Sammy Davis Jr, as the genre continued to evolve throughout the 20th century.

Key artists

Miles Davis
Louis Armstrong
Ella Fitzgerald
Duke Ellington
Billie Holiday

Key drummers

Buddy Rich
Gene Krupa
Eric Delaney
Tony Williams
Philly Joe Jones
Art Blakey
Elvin Jones

DOI:10.4324/9781003585169-7

Groove 35: Jazz example one

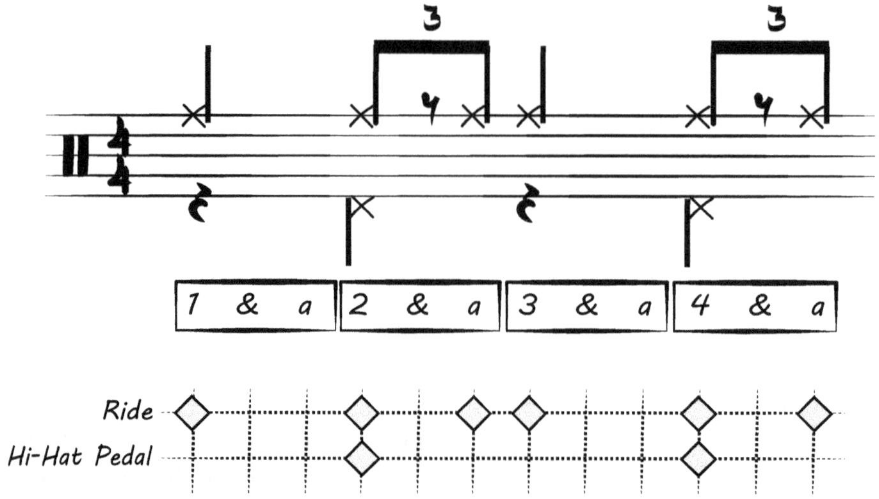

In this example we see the foundation for a lot of Jazz grooves. We have our Jazz ride cymbal pattern and the hi-hat pedal being played on beats two and four.

Groove 36: Jazz example two

In this second example, we have added the bass drum. This is played on the quarter notes (all four beats of the bar). Try a nice big boomy bass drum sound to help it feel a little bit more authentic within this genre.

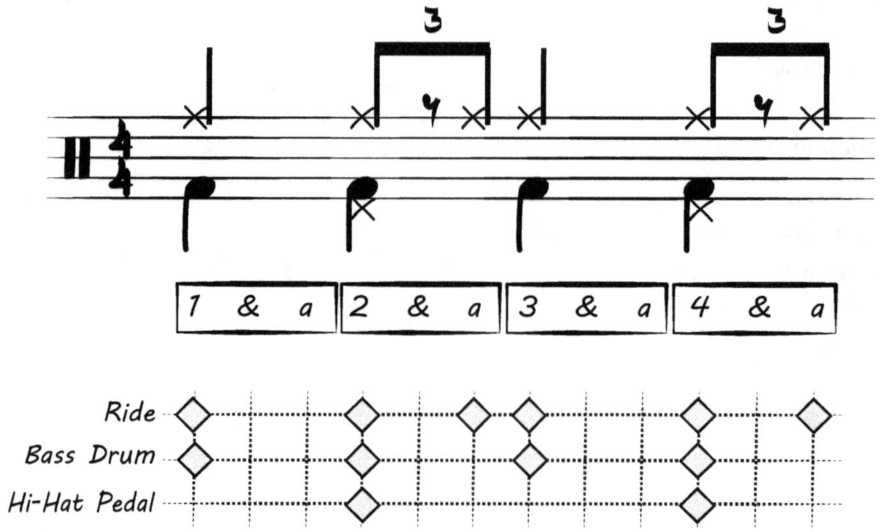

Groove 37: Comping example one

This is our first example of adding to the jazz ride pattern. Essentially, this is when the Jazz ride pattern stays the same, but the snare gets added on quarter notes.

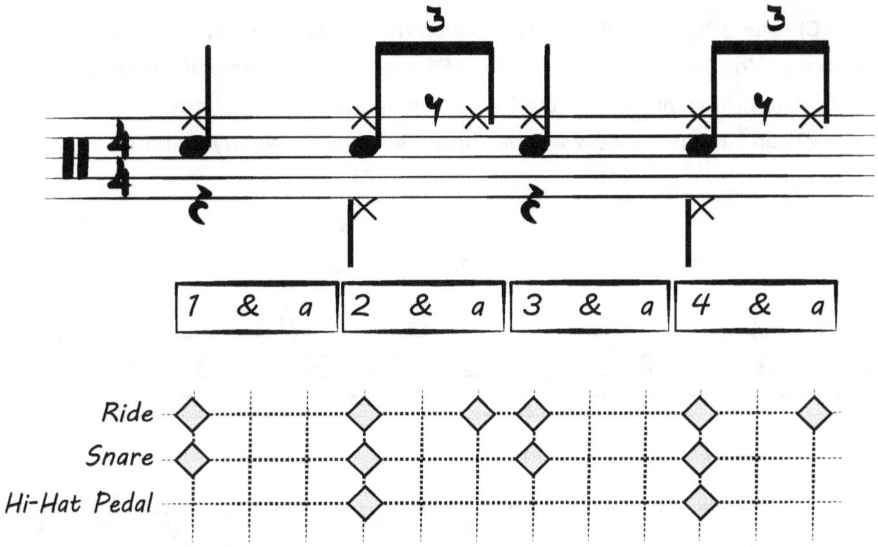

Groove 38: Comping example two

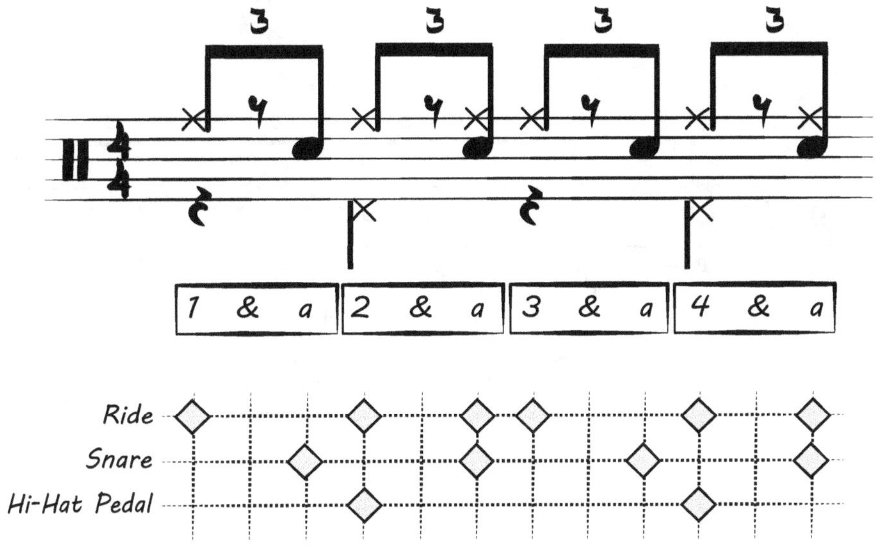

What we have here is our second variation. The previous example showed the snare on all four beats; now all of the hits are on the off-beats. Try doing this with the bass drum too, for a different effect.

From here you should be able to mix and match between these two examples and vary the snare drum pattern to create your own examples.

Groove creation: Challenge seven

Varying these examples

Keeping the Jazz ride cymbal the same (with the hi-hat pedal on beats two and four), there are more examples and ways in which we may utilise the snare drum to 'comp'. Books such as Gary Chester's *New Breed* and *Syncopation for The Modern Drummer* by Ted Reed are great for this. You can take a phrase, a line of notation or even the entire page and use these as rhythms to put on the snare drum (whilst keeping the jazz ride pattern).

Below is an example of how we may take a single line melody and then apply it to the jazz ride pattern.

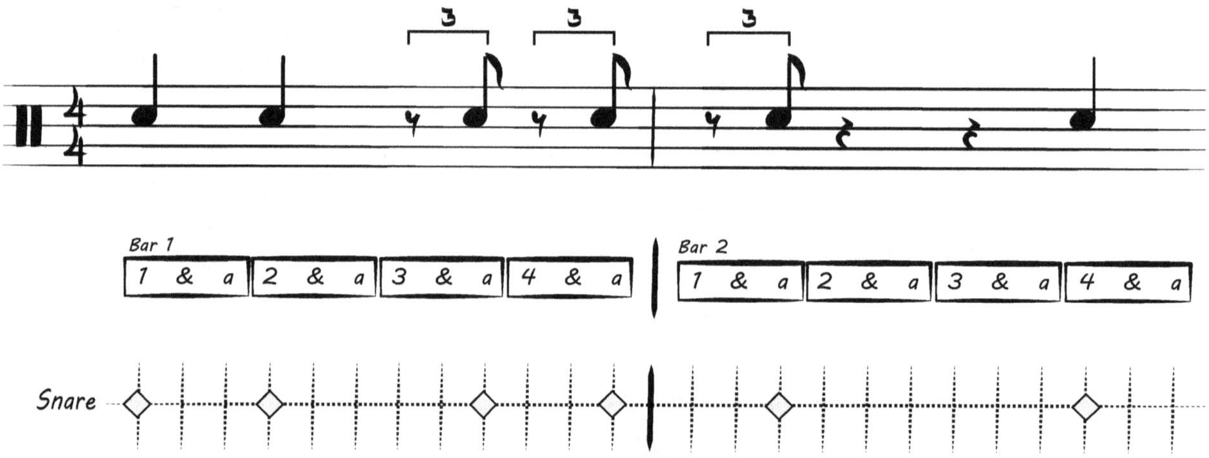

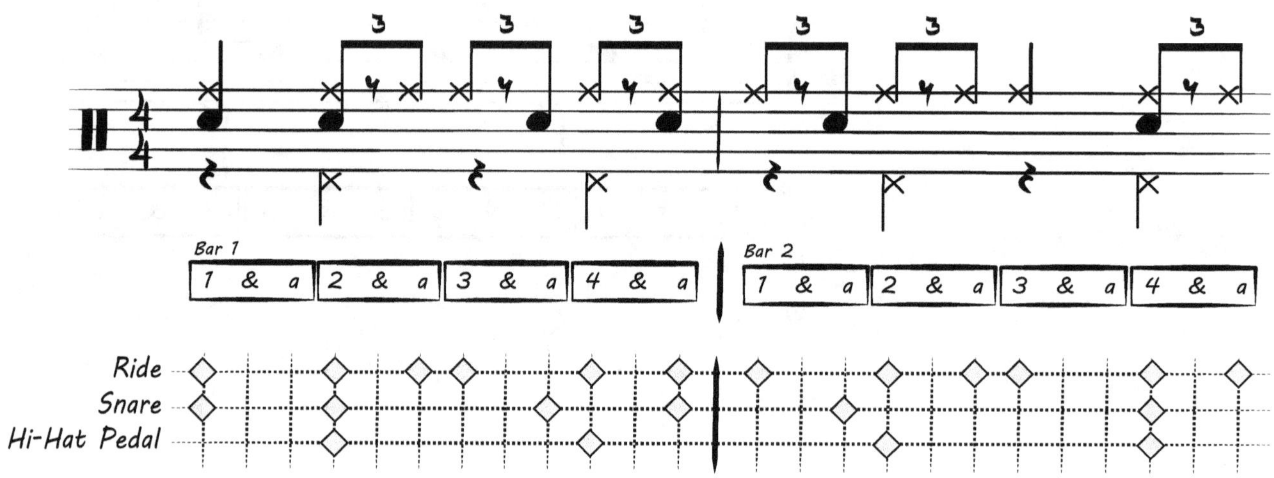

Orchestration of comping

All of the examples above can be placed and played on different parts of the drum set. For some variation of the above grooves, keep the Jazz ride pattern the same but try playing the melody on the bass drum. This allows you a completely new set of grooves and rhythms to give variety to the music.

With the example shown, keep the Jazz ride pattern consistent and try these variations of orchestration:

- Play melody on bass drum.
- Play any 8th notes on snare and any quarter notes on bass drum.
- Play any 8th notes on bass drum and any quarter notes on snare.
- Play any 8th notes as a cross-stick and any quarter notes on bass drum.
- Play any 8th notes on the bass drum and any quarter notes as a cross-stick.
- Any notes on '4+', play as an accent.

Experiment with this list. It will give you countless variations into this groove and what possibilities may exist under a Jazz ride cymbal pattern.

A note on this genre

At this point I feel obligated to point out that with this genre (and indeed with many others in this book), the grooves presented here are not the only defining grooves and musical examples available to us. For instance, in the case off Jazz, there is Big Band Swing, playing with brushes and indeed understanding and using forms that we haven't yet approached or discussed. For a more in-depth look into truly understanding the mechanics of Jazz drumming, explore John Riley's *The Art of Bop Drumming*.

When	What	Tempo	Notes/Reflections

Chapter 8 Rock, Punk and Metal

Genre overview

Genre introduction to Rock

The term Rock Music is very loose as it has evolved so much and there are so many subgenres. Throwing Rock, Metal and Punk together into the same chapter will inevitably upset some people, but, since they were closely interconnected, and I only have so many chapters, I decided to go with it.

Growing out of Blues and then Early Rock 'n' Roll in the 1950s, society saw the growing cultural importance of the teenager. The 'Teddy boys' then adopted this music, which then led, in the 1960s, to the so-called 'British Invasion' of the USA and bands such as The Beatles and The Rolling Stones. The Beatles frequently referenced their indebtedness to Buddy Holly and The Stones cited key Blues artists such as Muddy Waters as being among their key influences. As the 1960s, 1970s and 1980s progressed, more subgenres emerged, such as psychedelic rock, classic rock, hard rock, progressive rock and many more. Each of them had their own unique sound, bands and following.

Genre introduction to Punk

Emerging in the mid-1970s, Punk music expressed rebellion, pushing back against establishment norms. In the UK at the time (and also into the next decade), there was mass unemployment in major cities and recording artists were beginning to steer away from the mainstream culture of classic rock. Punk emerged with fast tempo songs, distorted guitars, raw energy and, possibly most importantly, attitude. At the same time, in the US, there were Punk glam bands such as the New York Dolls producing a fresh and unique sound. Indeed, many artists cite the New York Dolls as among their key influences.

Genre introduction to Metal

The heavier cousin to Rock is Metal. Now, this may be a section that you might feel you want to skip past if this is not your thing. But there is always something to be gained from any genre of music. Metal has a number of distinctive characteristics: the speed with which it is played, and its use of the double-kick pedal, drop-tuned guitars and growling vocals.

DOI: 10.4324/9781003585169-8

Other various subgenres of Rock, Punk and Metal music

- Punk Rock
- Post-Punk
- Symphonic Rock & Metal
- Hardcore
- Progressive Rock & Metal

Key Rock artists
The Rolling Stones
The Beatles
Led Zeppelin
AC/DC

Literally, there are too many to list here.

Key Punk artists
The Sex Pistols
The Clash
The Ramones

Key Metal artists
Black Sabbath
Metallica
Iron Maiden
Judas Priest

Again, so many more could be named, because of huge number of Rock and Metal subgenres.

Key Rock and Metal drummers
Bill Ward
Lars Ulrich
Joey Jordison
Dave Lombardo

Groove 39: Rock groove example one

This one took me a while to pick as there are literally so many within this genre. Here is one that is quite a staple rock groove and gives us a tremendous driving force with any song.

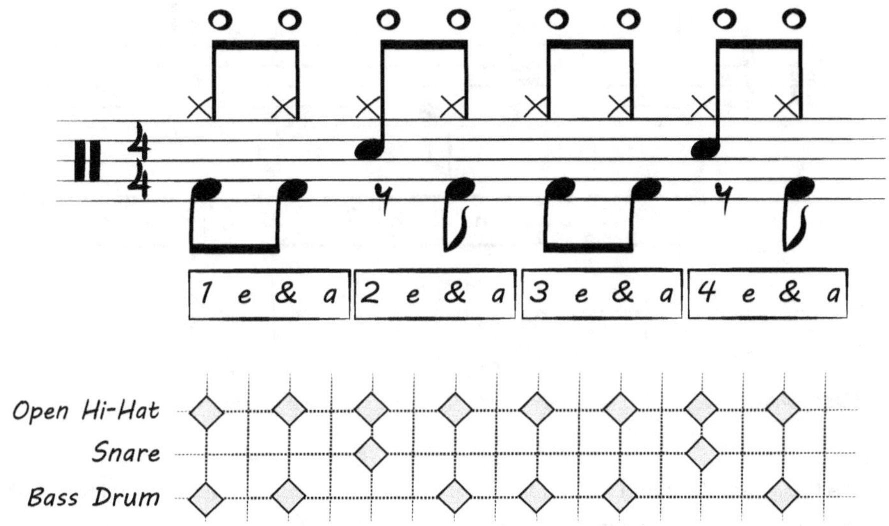

NB Interestingly enough, a similar groove was also played by Al Jackson Jr on My Baby Specializes. *So, in essence, we are taking an old soul groove, altering the tone of the drums and adding some loose hi-hats to add some 'grit', which relates to the overall drive and heaviness of this genre.*

Groove 40: Rock groove example two

What would a Rock section be without some cowbell action? This is a very 1980s hair metal American rock vibe groove. Among the bands that would often use this style of groove are Guns N Roses and Twisted Sister. This groove would also work well when played on the bell of the ride cymbal instead of the cowbell.

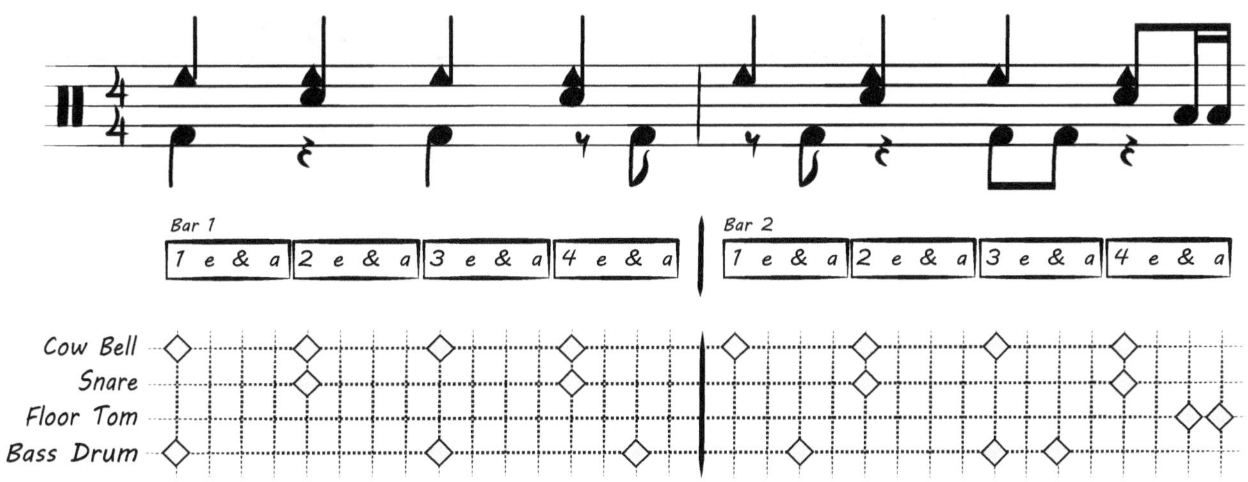

Groove 41: Punk example one

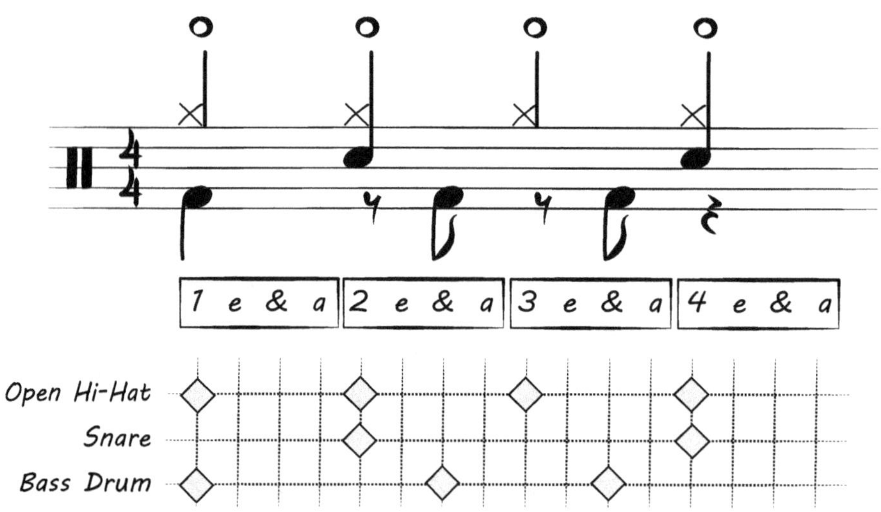

To make this example sound authentic, use it at a faster tempo and add some loose hi-hats. You are now on the way to creating an attitude-filled Punk track.

Groove 42: Punk example two

Here we have our second example of a Punk groove. Again, to get inside of this genre among the bands to explore are The Sex Pistols, The Clash and The Ramones.

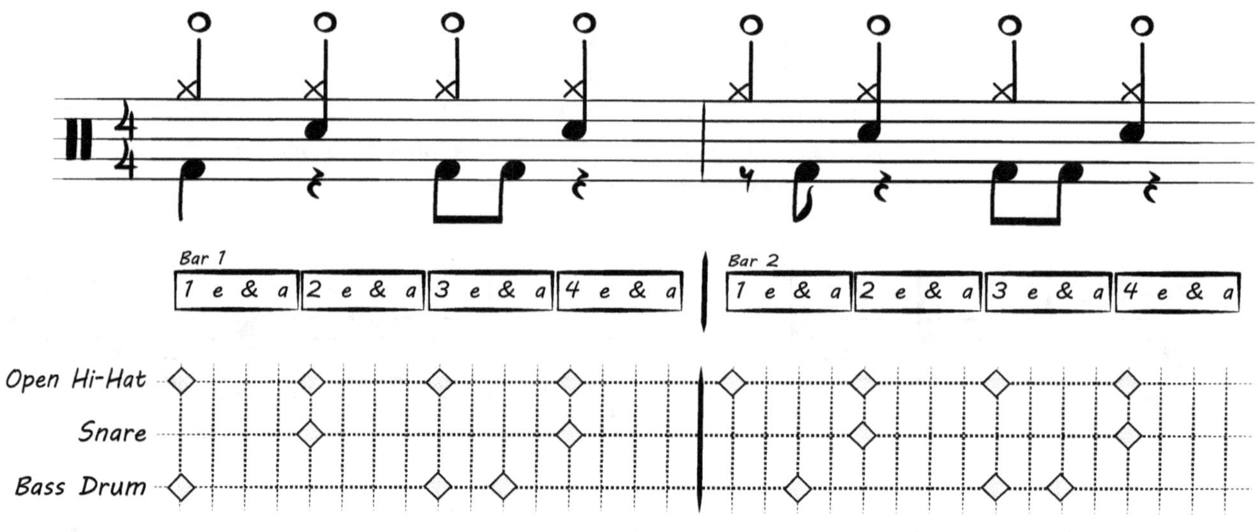

Groove 43: 16th note kick pattern

Now we start to venture into what may typically be found in Metal music. These examples are often used when drummers utilise their double kick drum pedal at faster tempos. Between these two patterns, the first one utilises 16th notes and the second one triplets. This gives us two different grooves with different feels.

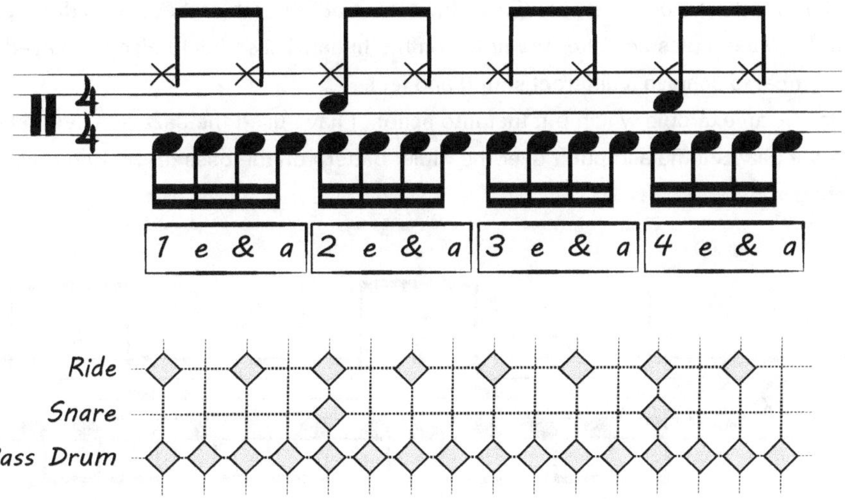

Groove 44: Triplet kick drum pattern

In this example, the bass drum is playing triplets. This gives a different feel to the previous groove. Both can be used effectively within Metal music.

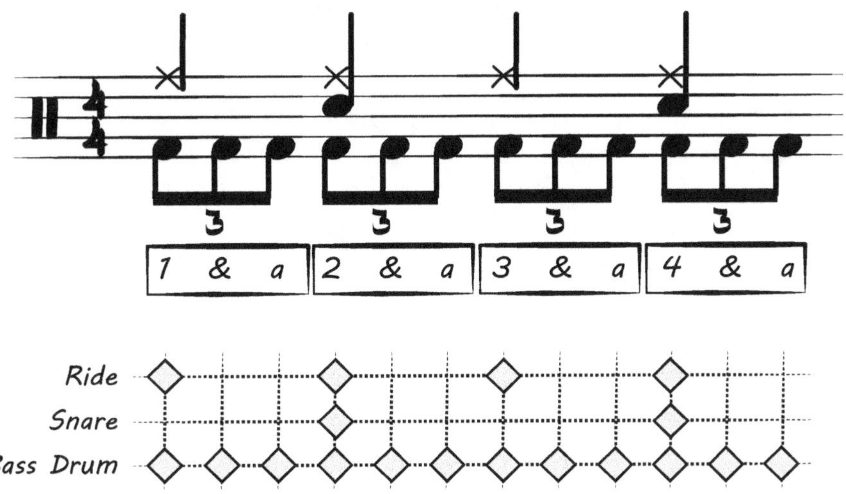

Groove creation: Challenge eight

With the previous two metal examples we can see, in example one, 16th notes on the kick and, in example, two triplets on the kick. To help create our own grooves within this genre, we keep the kick and snare patterns and then start to add our hi-hat options from Chapter 1 for the first groove.

Then (because we are in triplets) our hi-hat options from Chapter 2 'The Blues and Shuffles' for the second example. This allows us to utilise both of these bass drum patterns from these examples and alter the notes on the hi-hat. These could also be played on the ride cymbal to achieve a different sound and feel.

Here is an example which put that into action. I have used the Jazz ride cymbal pattern from our Jazz Shuffle and put it over the triplet pattern on the bass drum. Remember to try creating your own.

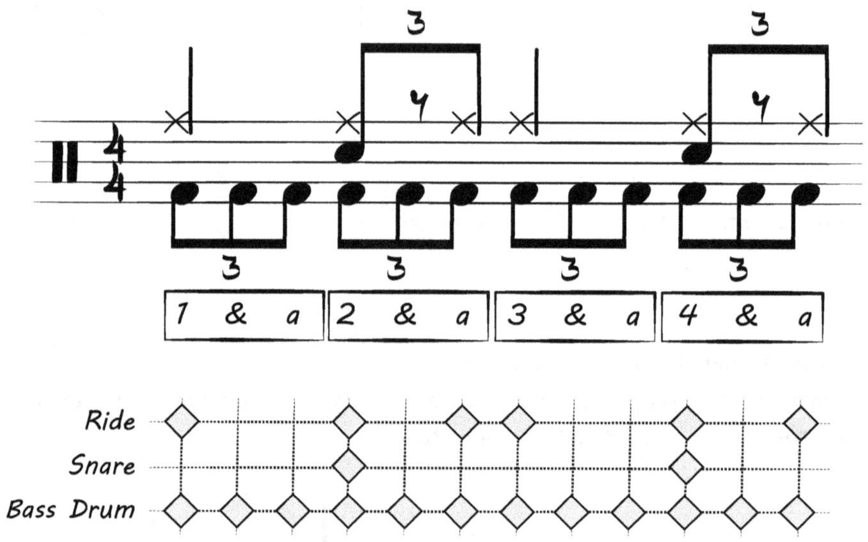

When	What	Tempo	Notes/Reflections

Chapter 9 Different time signatures

Introduction

Previous chapters have opened with an introduction to individual genres, each with odd time signatures. I don't truly believe this belongs to one genre exclusively, however. It is more of a concept that may be applied to any musical situation. Understanding time signatures and how they work is important for anyone involved in music creation, whether you are a drummer, producer or beatmaker.

Key artists

There are some key artists who use odd time signatures within their music and to great effect. Here are some examples (among many) to explore and check out:

Rush
Dream Theater
Genesis
Tool
Radiohead

Key drummers

Neil Peart
Mike Portnoy
Phil Collins
Danny Carey
Phil Selway

Counting and understanding time signatures

Time signatures are comprised of two numbers. The top number is how high we count; the bottom number is the rate of note we count (i.e. quarters, eighths etc). Thus, a bar of 7/4 literally means seven quarter notes, whereas 7/8 means we count seven eighth notes.

DOI:10.4324/9781003585169-9

Groove 45: The Waltz (3/4)

Arguably, I could have put this in the Jazz section. However, I felt that it was quite a nice bridge between these two chapters in the book. Here is our pattern in 3/4, which utilises a Jazz ride cymbal pattern. You could, of course, try some different patterns with the ride, snare and bass drum for variation.

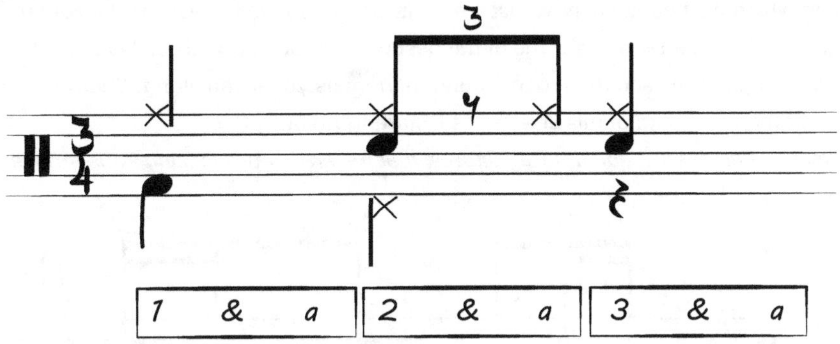

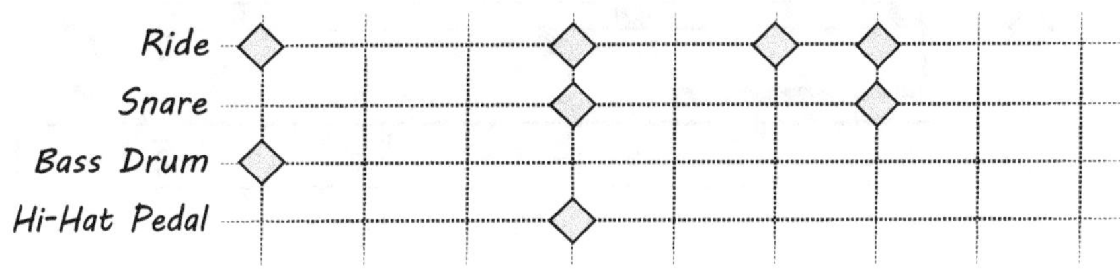

Groove 46: 5/4

Using the information on understanding time signatures, this groove is written in 5/4, which (as described) literally means we have five quarter notes. In a similar way to the Waltz, both variation and variety can be added to this groove.

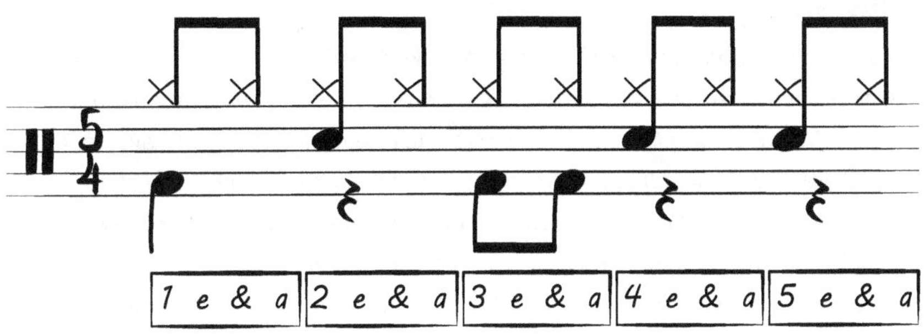

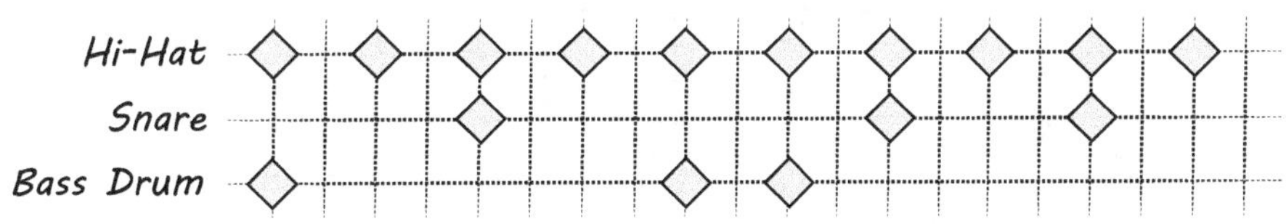

Groove 47: 5/8

Playing in anything over 8 (/8) can sometimes prove confusing when it comes to how to count. I see a lot of uncertain artists, musicians and producers who confuse time signatures such as 5/8 and 5/4 or 7/8 and 7/4.

The key here is understanding the pulse. Remember, the bottom number refers to the rate at which we count, i.e. how fast we do it. So, in 5/8 we are essentially counting five 8th notes (not five quarter notes). One of the best ways to develop this ability is to find songs in odd time signatures and then to try counting them as you listen along. Check out the artists listed at the beginning of this chapter as a starting point.

Remember, even in odd time signatures, that we are always looking for is groove and feel.

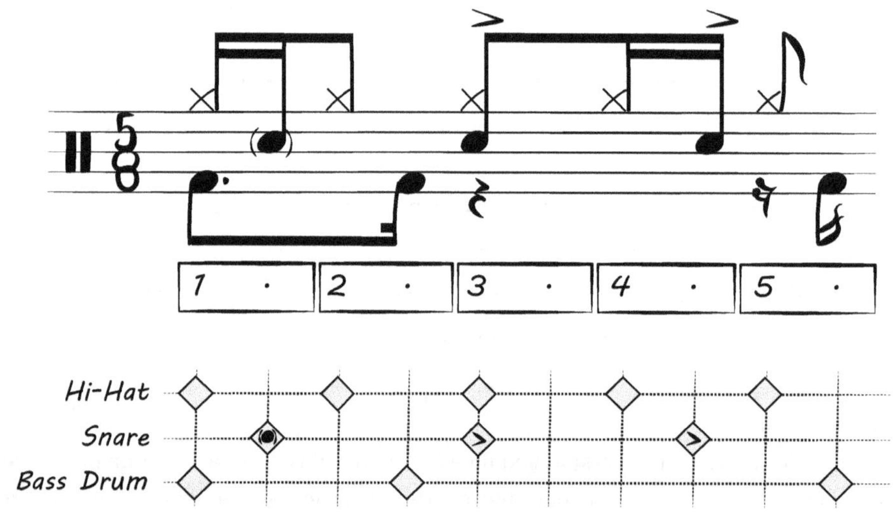

Groove 48: 7/8

One popular 'odd' time signature in which to make music in is 7/8. For instance, I play drums and percussion as part of a Progressive Rock group, Also Eden. Some of our songs, such as *Between The Lines* and *For Bumble*, as well as one on an upcoming album, all utilise this time signature. These are all examples of how a classic 7/8 groove may look.

NB It is very similar to playing in 4/4, except we shave the last 8th note of each bar off.

Also, when counting this aloud, don't say 'sev-en'. This is a two-syllable word which can cause you to accidentally add an extra 8th note. This is why 'sev' is written.

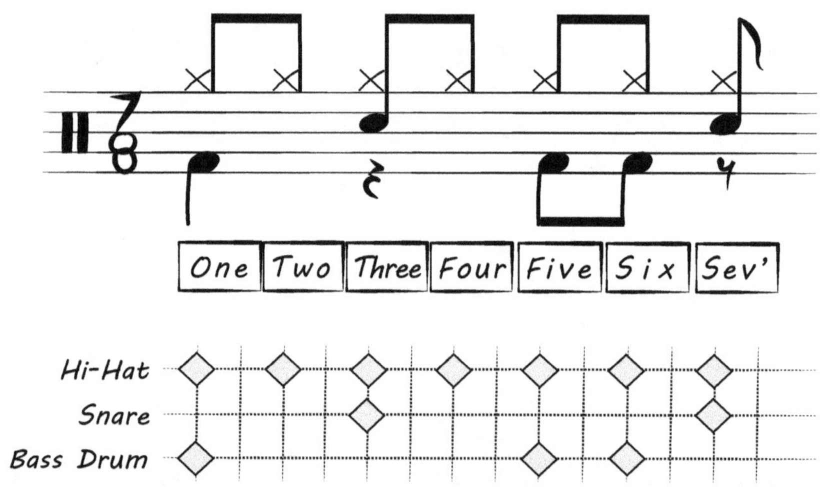

Groove 49: Funk style 7/8

What I have done here is take the previous example of playing in 7/8 groove and 'Funked it up' a bit by using some of our previously discussed Funk characteristics. Note the open hi-hats and additional ghost notes which we have come across previously.

I could go into greater depth here about how we can take different time signatures and 'make them Funky'. You are your own experimentation into grooves; take a new time signature and explore adding some of the Funk characteristics, as described above.

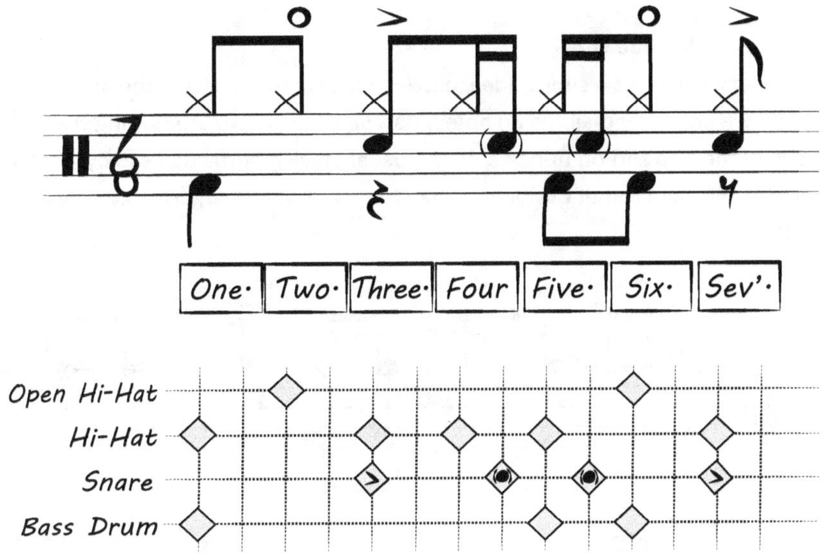

Groove 50: Quarter notes over the bar line feel

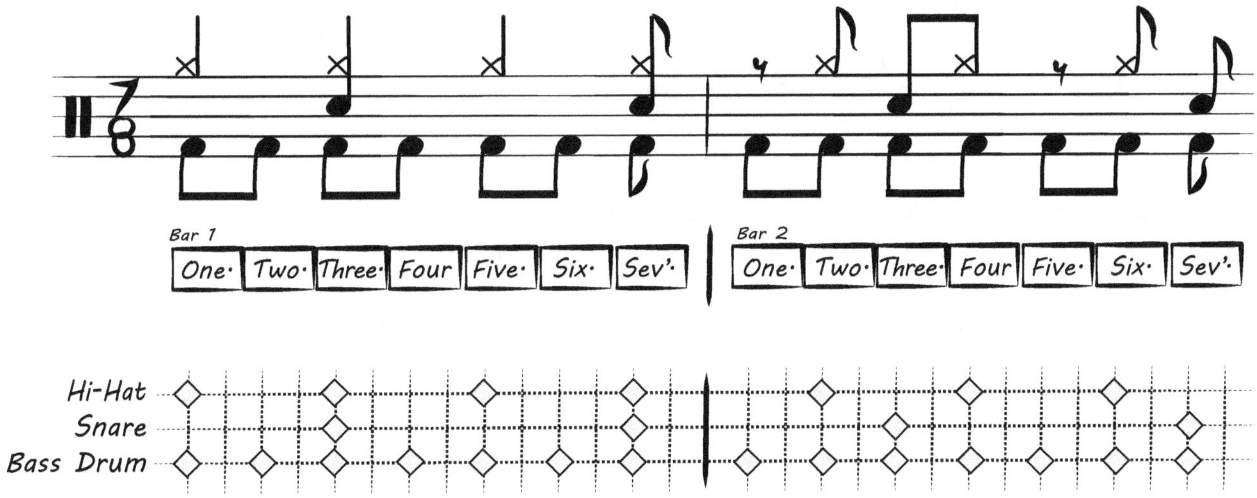

Here is a cool little thing you can do to 'trick' your audience. This is a two-bar phrase in 7/8 with a quarter note hi-hat pulse. Because we are not in a 'over 4' (/4) time signature, this show the hi-hats switch to the offbeat on the second bar. Because it is a two-bar phrase, however, it still gives us a consistent quarter note pulse that repeats every two bars.

Equally, having the consistent quarter notes pulse will give the audience something to nod their heads to whilst you create music in a 7/8 time signature. Check out the example shown. You can take this idea and alter the notes played on the snare and bass drums whilst maintaining the placement of the hi-hat notes to create your own examples in this style.

Groove 51: Over the ride line

In this groove, we explore a similar idea, except we have moved from the hi-hats to the ride cymbal and we are playing all the 8th notes. As you can see, we are putting the downbeats on the bell of the ride and on upbeats on the usual playing surface of the cymbal. Because we are in 7/8, this means that this gets flipped for every other bar, creating a new and interesting effect.

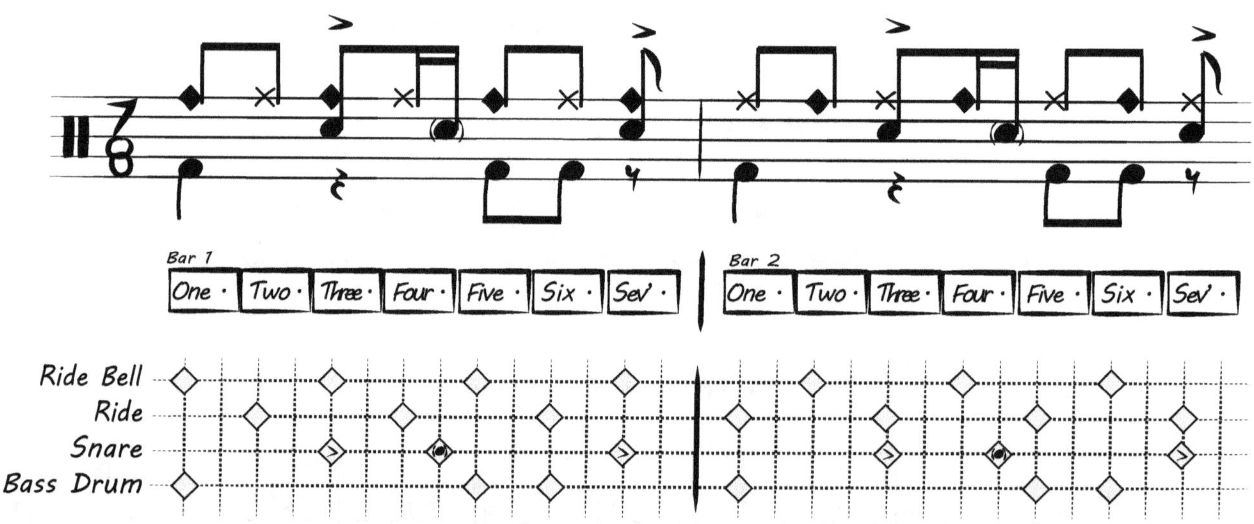

Groove 52: 11/16

For our last groove within this chapter, we look at playing in a time signature 'over 16' (/16). We are now approaching this by counting the 16th notes. Playing/programming over 16ths can feel a bit odd and staggered at first, but stay with it.

In terms of how I have created this groove, in this example I have split two paradiddles (R L R R L R L L) between the ride, hi-hats and snare. We have then added a bell and two notes on the tom tom at the end of the phrase. This is, quite literally, eleven 16th notes.

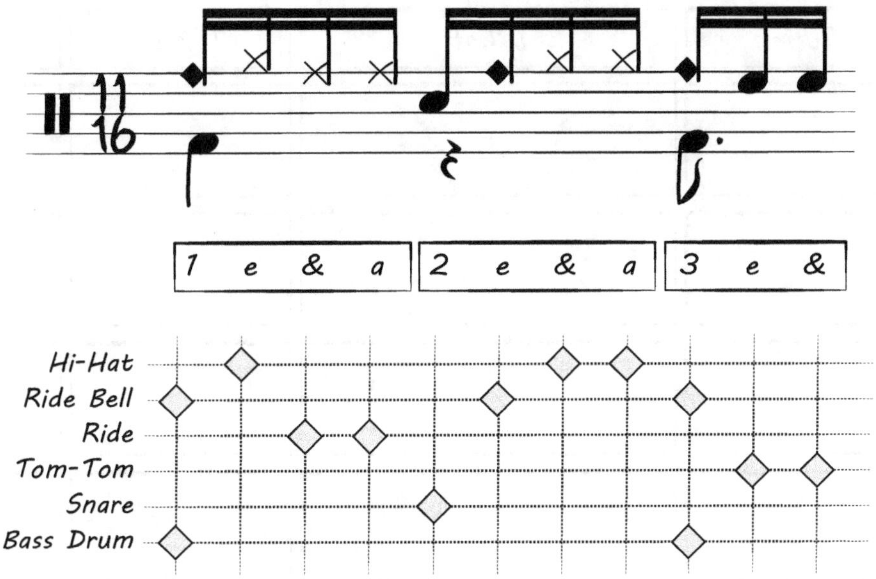

NB Another way to look at this, is that we are playing a bar of 4/4 but shaving off the last five 16th notes. For ease, I would count this as shown on the image '1 e + a' etc. rather than just counting up to eleven each time.

Groove creation: Challenge nine

The groove creation challenge for this chapter is to take over the bar line concepts of groove and try it in different time signatures.

Because of the alternating 8th note feel between bars, this will only work with time signatures over eight (/8). Here is an example of taking the concept from groove 51 and applying it to 5/8. Experiment with this in 9/8, 11/8, and so on. In addition, try some different bass drum patterns and hi-hat options.

There are some examples of these concepts in 5/8 to get you started.

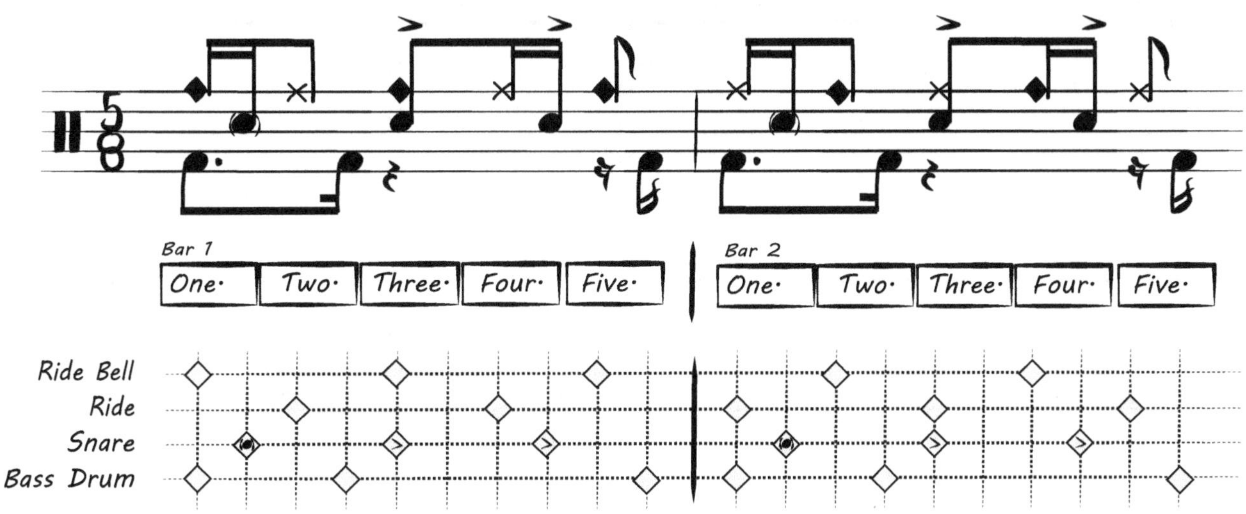

When	What	Tempo	Notes/Reflections

Chapter 10 Different cultural grooves and rhythms

Overview

Normally at the start of these of the chapters I give a little discussion about the genre of music and some key artists. In this chapter, however, I think it is important to acknowledge that we are only scratching the surface of different cultures. By this I mean that it goes beyond examining the music to consider the culture as a whole, and the integral role which music may play. From a brief glance at some of these different cultural grooves, we are able to look at how we can amalgamate them not only with each other but also with different grooves we have encountered so far to create something new.

The Claves

Something that can help us get inside of these challenging rhythms is understanding the Clave. When I first started to explore Latin music, I sometimes struggled to identify the first beat of each bar. An understanding of the Claves can help with this.

Firstly, the word Clave be applied to two things. The first is the actual instrument, two small handheld sticks, which are struck together to produce a high-pitched note. The second is that Clave is our rhythmic key for the music we are playing. Within Afro-Cuban music, we have a range of separate styles and dances, including Rumba, Son and Salsa. Below is a guide to these Clave rhythms. This can guide us through the music and the drums we may be based around these Claves to help give the music the feel and vibe we are trying to create. More of this will be shown in this chapter.

Son Clave

The first example we have is the 2:3 Son Clave. This essentially means tht the first two beats are played in the first bar and the final three beats in the second bar.

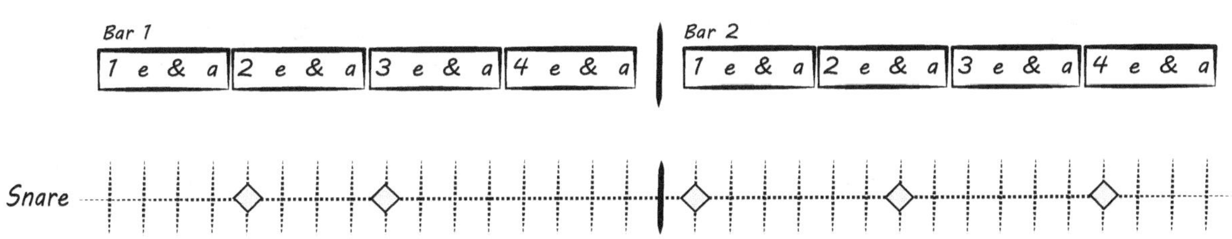

DOI:10.4324/9781003585169-10

Equally, if we take these two bars and switch them around, we have the 3:2 Son Clave, as shown below. If you refer back to the Bo Diddley beat in Chapter 3, you can hear this in action in a musical context.

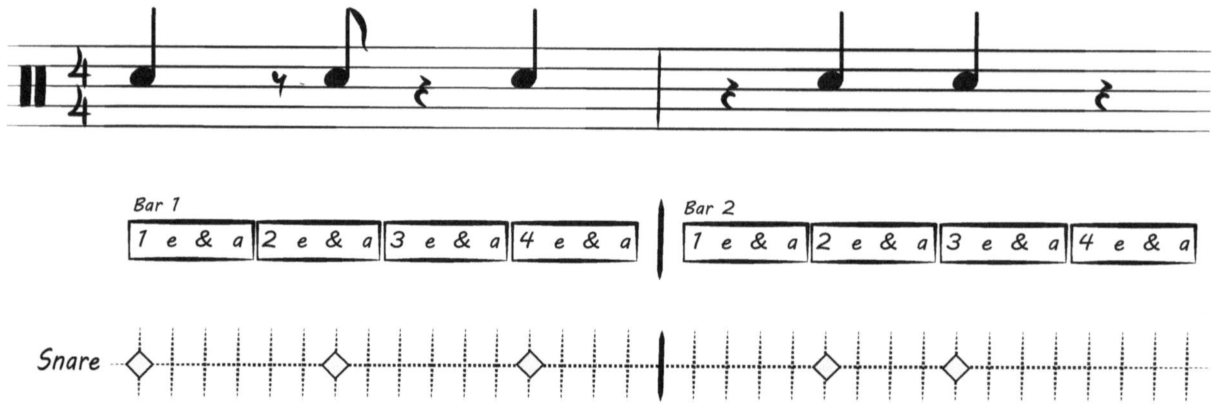

Rumba Clave

Next we have our Rumba Clave, which differs slightly. In the 3:2 version, the Rumba is identical to the Son, but we are just shifting beat four over by an 8th note to the 'and' of beat four. This is displayed below.

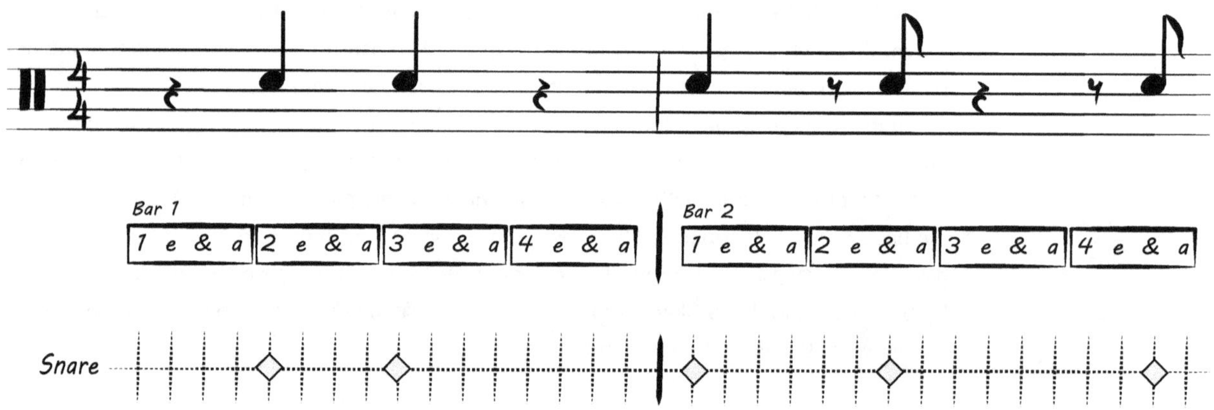

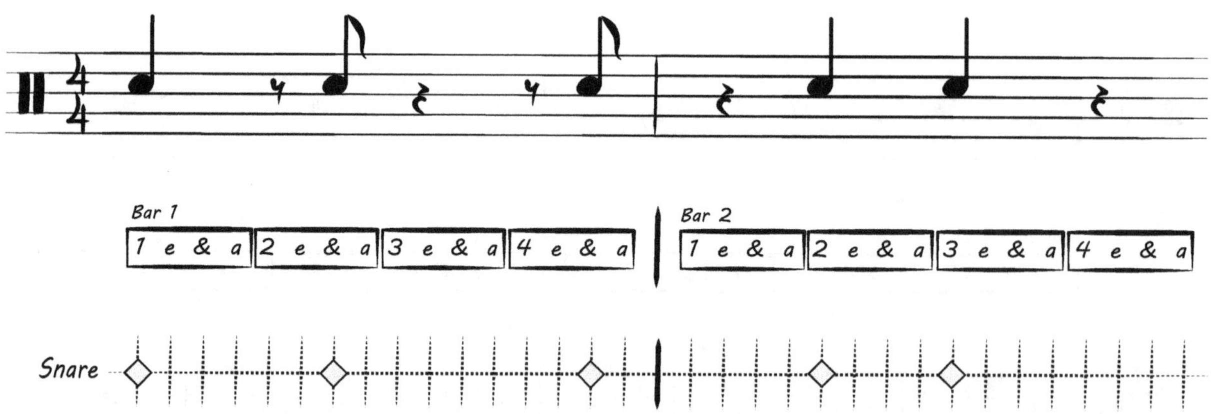

Afro-Cuban

In a nutshell, Afro-Cuban music essentially combines rhythms and melodies inspired by the music of Africa and Cuba. This draws on the historical impact of the transatlantic slave trade, when millions of Africans were transported from the continent to the Americas. This African influence merged with the European influences it encountered (particularly those from the Iberian peninsula) and this blend of cultures, rhythms and styles helped birth Afro-Cuban music.

We are just scratching the surface of the music here, but it's a good starting point. Here I was conscious to stay away from the term 'world music' as I often find that it implicitly suggests that English and American music is dominant and that the rest of the world just gets clumped together in one single genre. Such a classification that feels like a disservice to the cultures and histories that go into this music.

It is also important to remember that many of the rhythms we are playing and replicating here were not always originally designed for the drum set. Originally, they had been played on different pieces of percussion. The other interesting point for producers is that, because of the different percussion samples, each of the limbs from the grooves below could be played on any assortment of percussion that you choose.

Groove 53: The 3:2 Cascara

This rhythm was originally played on the timbales. *Cascara* is Spanish for shell as this rhythm was originally played on the shells of the timbales. This groove can be split between two sound surfaces; with a drum set, it is often the cross-stick plays the Clave pattern. This can become varied as you become more comfortable with this groove and style.

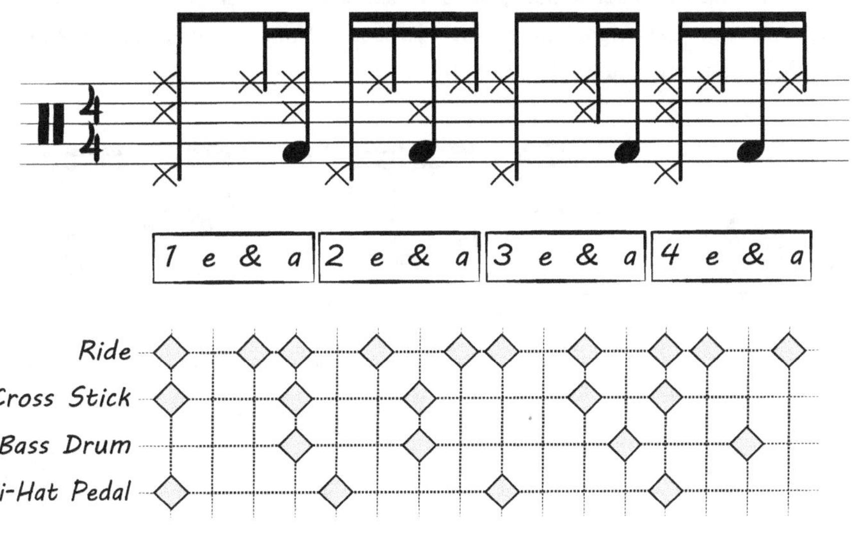

NB In a similar way to the Clave rhythms described, these two bars can be reversed (so that you play the second bar first and the first bar second) to create a 2:3 cascara pattern.

Accents in Latin music

Sometimes, the hardest thing about these grooves is lifting them off the page. Adopt this approach; whenever you have two notes together on the main timekeeper (ride, hi-hats, cowbell etc), try accenting the second of every two consecutive notes. First trying playing/programming this without the accents; then do it with the added accents. Notice the difference, and see how it can help lift a groove off the page and bring it alive with feel.

Groove 54: The Songo

The Songo is the first widely known Afro-Cuban groove to be created for the drum set. Note the tumbao foot pattern on the bass drum. The hi-hat can pedal can be utilised when this groove is created on the ride cymbal or cowbell.

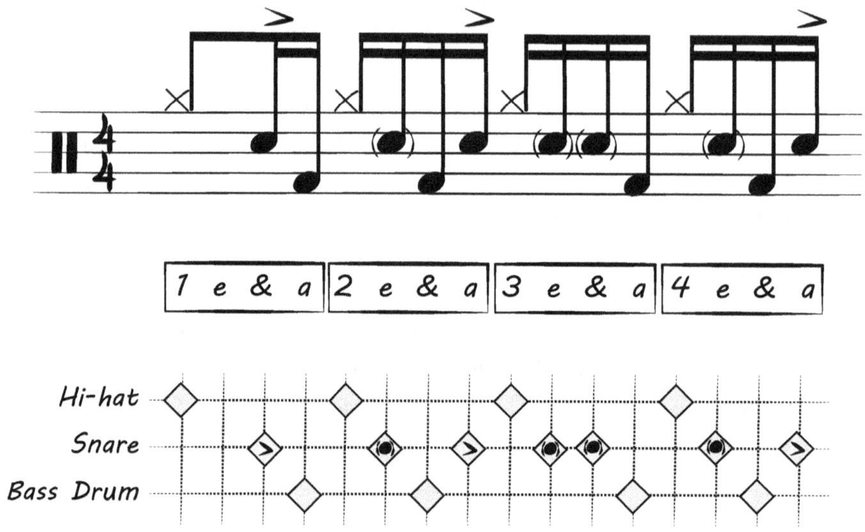

Groove 55: The guaguanco

Okay, first things first; its pronounced *'wah-wan-co'*. This refers to the melody between the floor tom and tom-tom used in this groove. This is as bit of an amalgamated groove of what we have seen so far. The right hand is playing a cascara pattern; the left plays the guaguanco melody and the feet play a tumbao pattern.

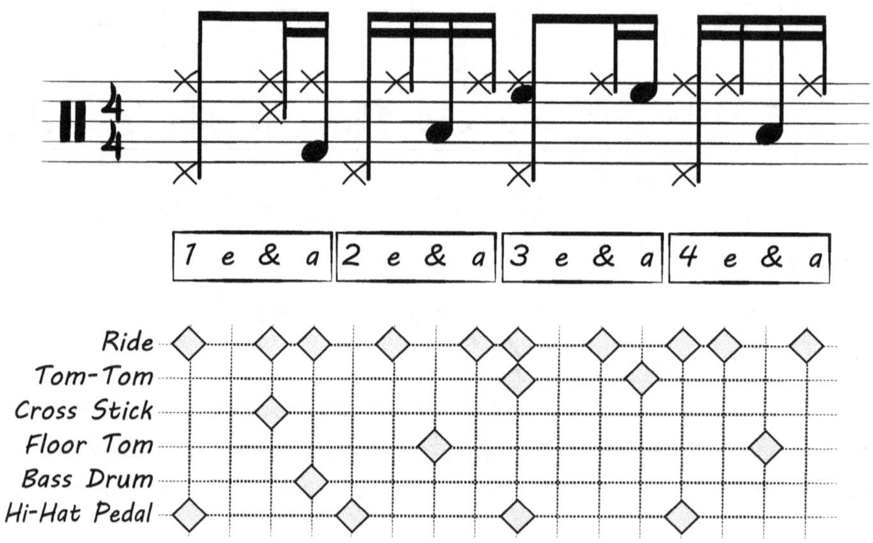

Brazilian

Overview

Brazilian music is not quite like any other. Known for its carnival and dance, Brazilian can be split into many genres, such as Samba, Batucada (a subgenre of Samba), Bossa Nova, Choro and Forró. I highly recommend that you listen to some of these genres to start with to understand how it combines different instruments.

Percussion to drum set application

In a similar way to some of the Afro-Cuban grooves we have seen, a lot of the Brazilian grooves were originally played on different pieces of percussion. So, what we are doing here is taking rhythms from different percussive instruments and adapting them to the drum set. Of course, for producers/beatmakers some of the percussion heard in these genres can be programmed and used within your music.

Groove 56: The *partido-alto*

The *partido-alto*, which can be translated as 'high party' in Portuguese, can refer to the subgenre of Samba which comprises two singers battling verses in an improvised form. It is also widely known as a rhythm; hence its inclusion in this book. In this example the *partido-alto* rhythm is played on the cross-stick. The bass drum (replicating the Surdo drum) and the hi-hat pedal is playing what is known as the Samba Ostinato.

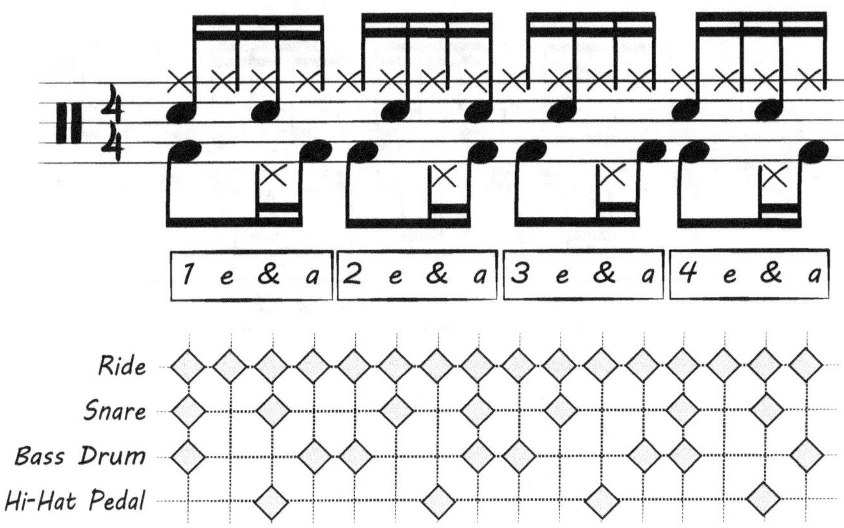

Groove 57: The Bossa Nova

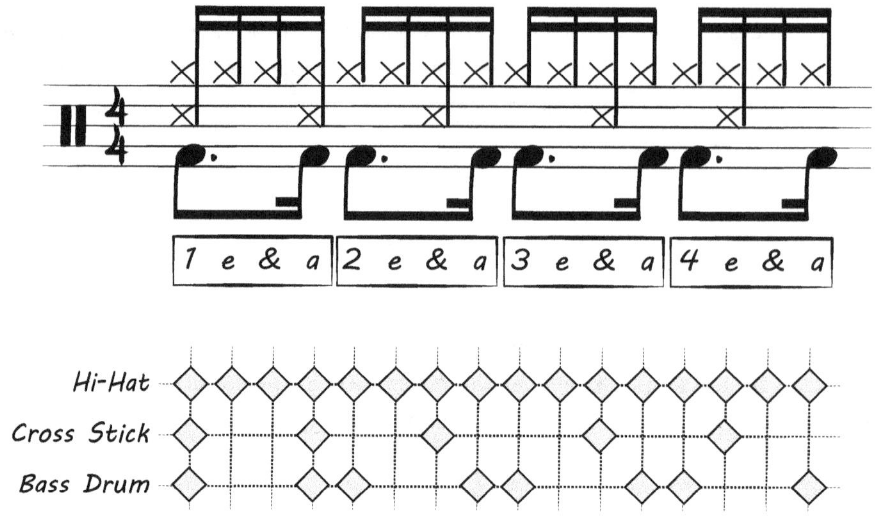

Sometimes simply referred to as a 'Bossa', this is a staple rhythm for the drum set in terms of the different cultural rhythms that are seen and heard within western music. Here I have written it using hi-hats, but this could be replaced with the ride cymbal and the hi-hat pedal played on two and four to give us that full Samba ostinato once again.

Groove 58: The Jazz Samba

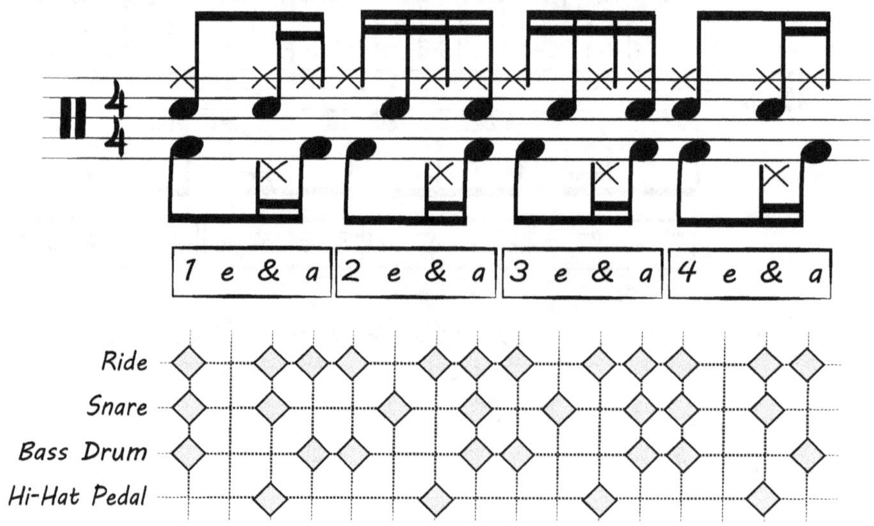

As mentioned at the beginning of this book, many of these grooves derive through rhythms played on pieces of percussion. This is an example of how Brazilian rhythms can be used in a groove that would work in the context of Jazz or Fusion. Follow the cross-stick pattern as shown but as you get more comfortable with it feel free to break away from it but try to keep the flavour.

Underlying rhythms

Often played between the bass drum and the hi-hat pedal, these three ostinatos can be used as foundation grooves for Latin and Brazilian drum set playing. These ostinatos can be used underneath some of the snare and ride patterns found in this chapter.

Baiao Ostinato

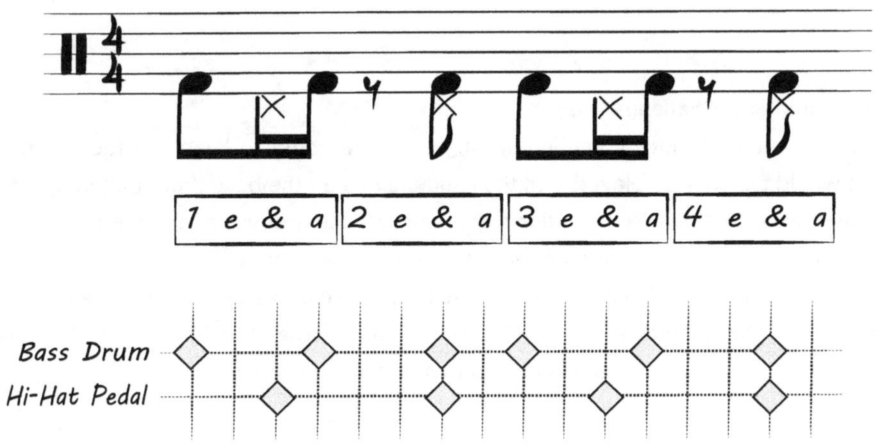

Tumbao Ostinato

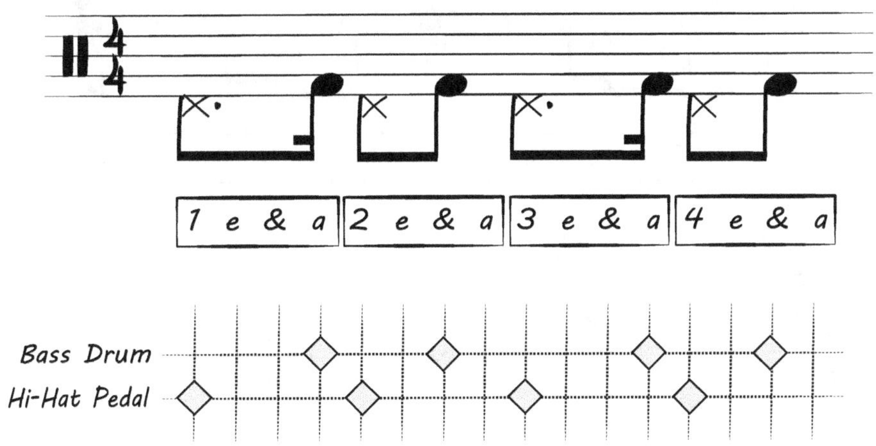

Drum Groove Creation

Samba Ostinato

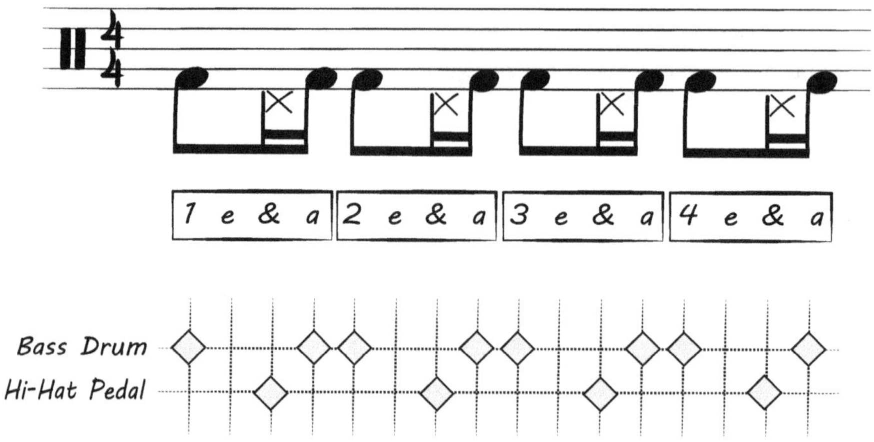

Groove creation: Challenge ten

For this chapter's groove creation challenge, we are going to keep our topline rhythms (what would typically be played with the hands) and alter the bass drum patterns to one of our underlying rhythms. You'll notice that many of these grooves already use one of them, so your challenge is to try the other ones to see how they sound.

Here is an example of this in action. I have taken the cascara pattern and guaguanco tom melody from this chapter and put a Baiao ostinato underneath it. Try this out for yourself and create your own using the underlying rhythms shown above.

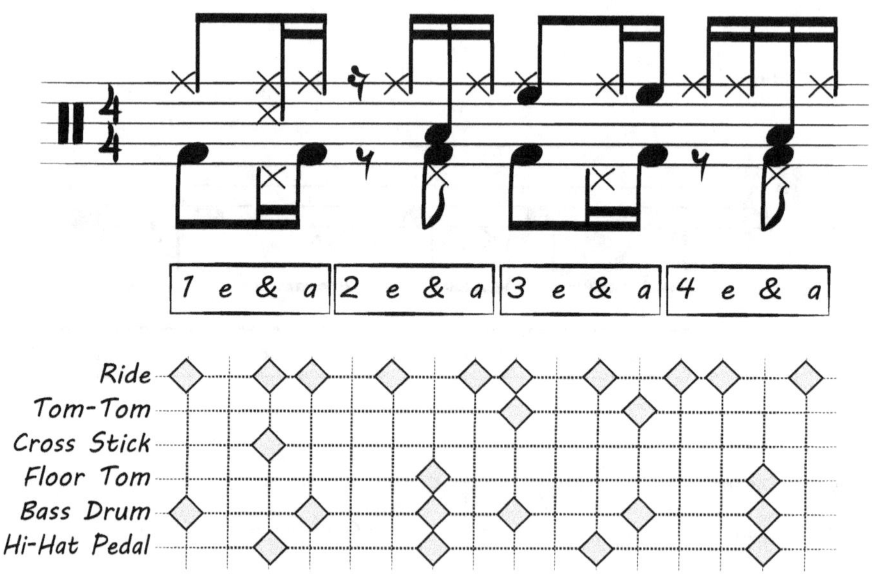

When	What	Tempo	Notes/Reflections

Chapter 11 Other 'must know' grooves

Introduction

Up to this point we have discussed a variety of genres, styles and different drum grooves found within different cultures and time periods. This could, and always will, be expanded upon as new styles and genres of music emerge. New artists are coming out with new beats and new drum grooves that help inspire us in our own drumming and music creation.

Due to the expansive nature of so many different genres and grooves, I wanted to pop in a chapter of some 'must know' grooves for you to be aware of. As with anything found in this book, check out the examples shown; more importantly, however, take it, experiment with it and make it your own.

Groove 59: Afrobeat

It feels a crime to reduce a whole genre to a singular groove. Afrobeat and the drumming of Tony Allen is something I would suggest exploring. Afrobeat can be traced to West Africa, and within it we can hear some great melodies being played on the drum set.

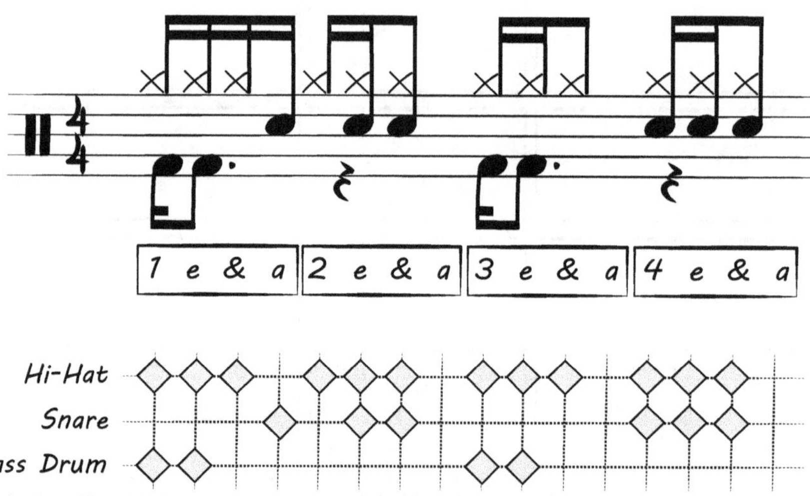

Groove 60: Modern Afrobeats

There can also be some confusion between the two terms Afrobeat and Afrobeats. They are separate entities with different sounds and operating within different genres. Afrobeat is a blend of African music and American soul and jazz influences.

DOI:10.4324/9781003585169-11

In contrast, Afrobeats draws on other influences, such as Ghanaian Hip-Hop. Two of the most prominent Afrobeats artists, and worth checking out, are Burna Boy and Wizkid. Afrobeats has a huge scene in the UK and is hugely popular amongst Black communities.

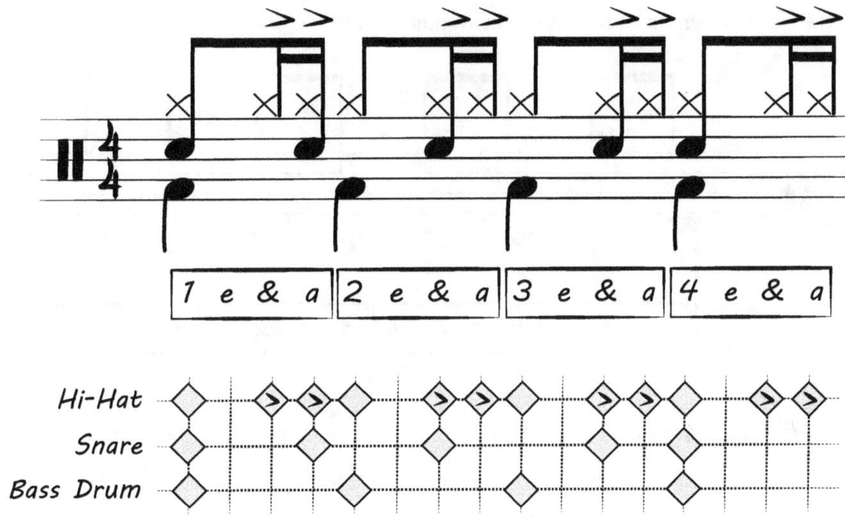

Groove 61: Double time feel

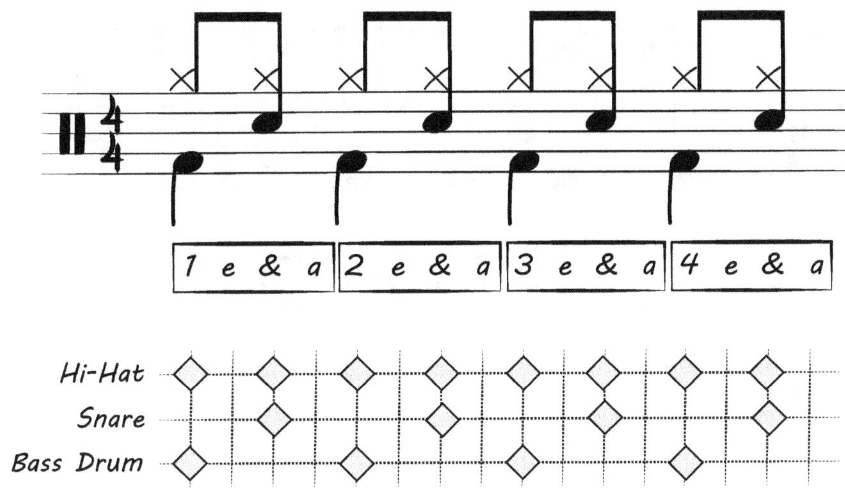

If you have come across the term metric modulation, this is what we may think of as implied metric modulation. Essentially, it sounds (to the listener) like you are getting quicker, when in reality you aren't. This may be achieved by altering the placement of the snare drum onto the off beats. Try moving between this example and the Groove one to hear the difference, and the effect this has.

NB This is similar to what we saw in our Punk Groove. The reason it is making an appearance here is that it shows the contrast between putting the snare on the off beats as opposed to beats two and four. To hear this in action listen to Should I Stay or Should I Go *by The Clash.*

Groove 62: Half time feel

Here we have the opposite effect. In this example of implied metric modulation, it feels (again, to the listener) like the beat is slowing down. Remember we are not *actually* slowing the groove down; we are just from moving the backbeat from two and four to beat three to get this different feel. Try going between these two examples and the *Money Beat* to see the effect that the placement of the snare drum has on a track.

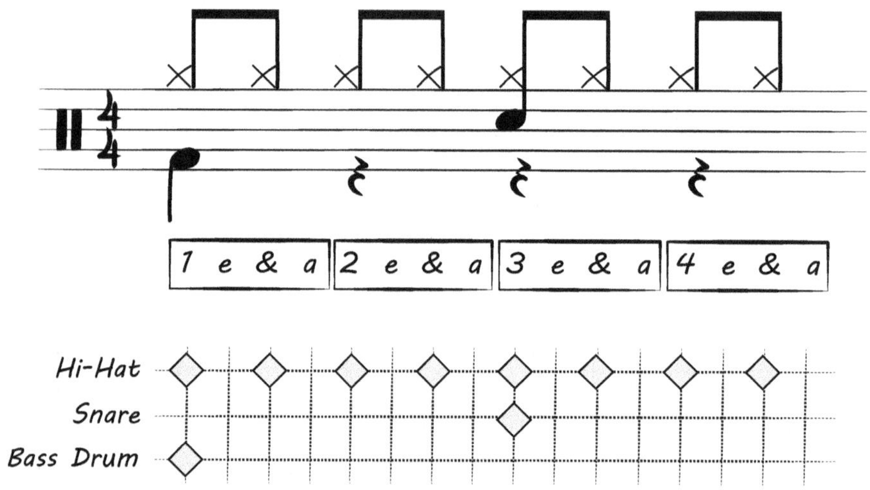

Groove 63: New Orleans Funk

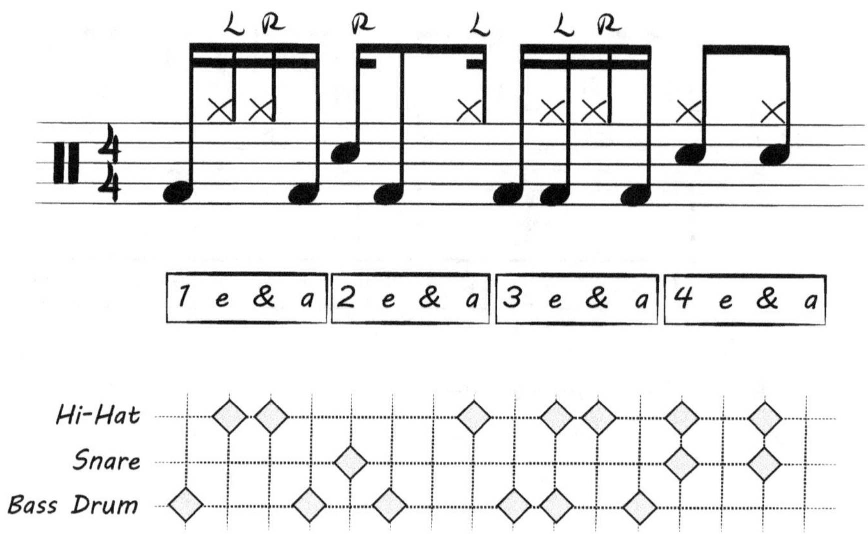

I originally had this one in the Funk chapter. But I wanted to put it as a stand-alone groove to emphasise how important I think it is.

This comes from the band The Meters. *Cissy Strut* is arguably the most famous groove that Zigaboo Modeliste is credited with. Many drummers notoriously misinterpret this and attempt to play it one-handed. This groove was played with two hands on the hi-hats and coming down for the backbeat with the right hand.

Drummers: please take careful note of the sticking above. You can hear some variations on the record, but this is the essence of the groove. Remember to try and get that in between straight and swung feel.

Producers: don't programme this without any swing or it will sound too robotic.

Groove 64: The Reggae one drop

A genre that often preaches love and is steeped heavily in campaigning for political justice and equality, Reggae developed on the island of Jamaica in the 1960s as it began to gain popularity around the world. This genre is closely associated with the Rastafarian religion and also the genres of Ska and Rock Steady. On the surface, this may feel like a very simple groove; making it sit just right, however, can completely transform a song. Stewart Copeland used this to great effect with The Police, proof that a groove can work outside of its original context.

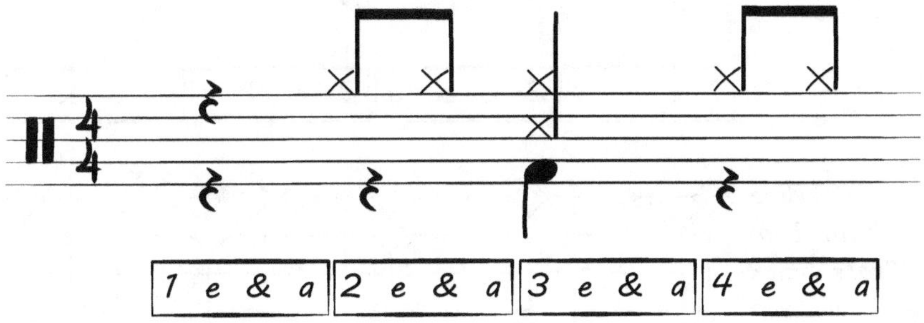

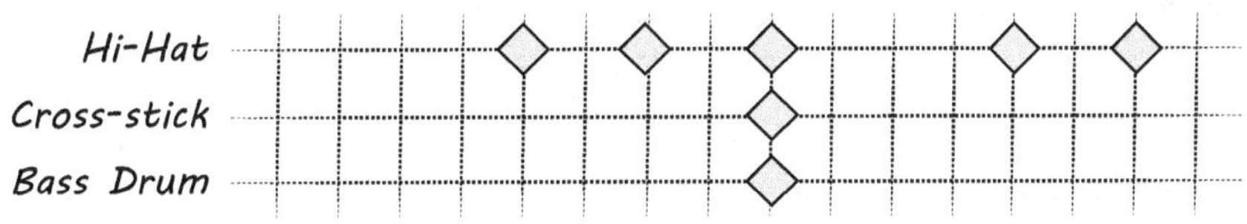

Groove 65: The Mozambique

Made famous by the legendary Steve Gadd, check out Paul Simon's *Late in The Evening* to hear this groove in action. This is just one variation of The Mozambique; experiment with putting *four on the floor* and different surfaces instead of the ride cymbal.

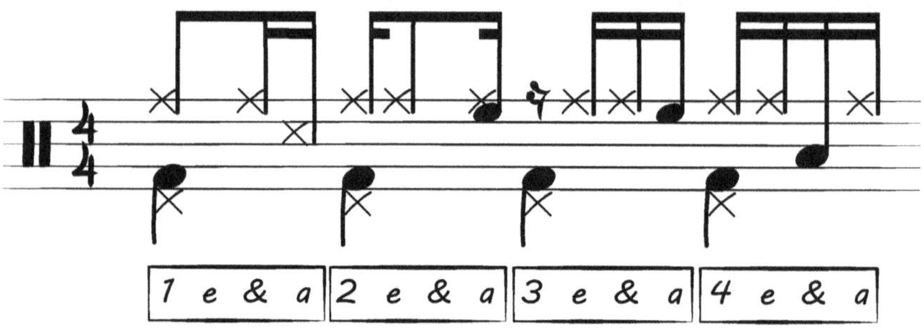

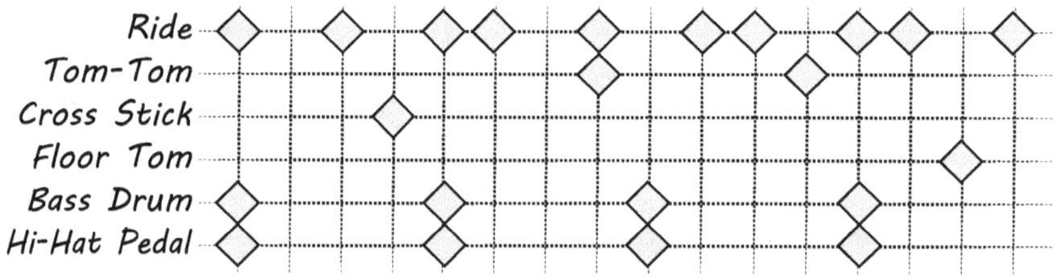

Groove creation: Challenge eleven

In this chapter, the challenge is based around your ability to play/programme these rhythms with your own manipulation. By this I mean: how can you take these grooves and vary them? Here are some things to try and experiment with…

- How would this be if you altered the tempo?
- How do these grooves sound if they are straight, swung, or somewhere in-between?
- How could the bass drum pattern be varied (maybe using 16th note grid)?
- Could ghost notes be used to add some additional texture?

Beyond this, you could use some of the other concepts previously discussed in this book (hi-hat patterns, Funky characteristics) to alter these grooves and create something that sounds new and unique.

When	What	Tempo	Notes/Reflections

Chapter 12 Creating linear drum grooves

Overview

In a nutshell, linear means that only one part of the drum set is being played at a time; we can find this style of groove in the playing of Tower of Power drummer David Garibaldi. Understanding and utilising linear grooves can help bring a new feel and flavour to our Drum Groove Creation.

I often find that people aren't taught the formula of how we can create linear grooves. Instead, drummers are inundated with clickbait titles such as 'must know linear groove or lick'. Whilst I understand the appeal of such posts, it limits us to what other people have created and memorising already created linear phrase, grooves and fills. Having our own method for devising linear grooves is key to unlocking our potential in Drum Groove Creation.

Let's explore the Linear Drumming Method...

A linear drumming method

1. Pick a time signature
2. Pick a subdivision
3. Use groupings to make up a bar
4. Orchestrate and find a backbeat

Groupings

This refers to a set amount of notes within a subdivision. For instance, generally in 4/4 we tend to group the 16th notes into groups of four. However, instead of simply just grouping into four we can do groups of other numbers. Below, for example, we have taken a cluster of three notes and placed an accent on the first of every three. We are still in 4/4, and we are still in a 16th note subdivision; however, we are now dividing this bar into groups of three. Check it out below.

DOI:10.4324/9781003585169-12

Groupings of three

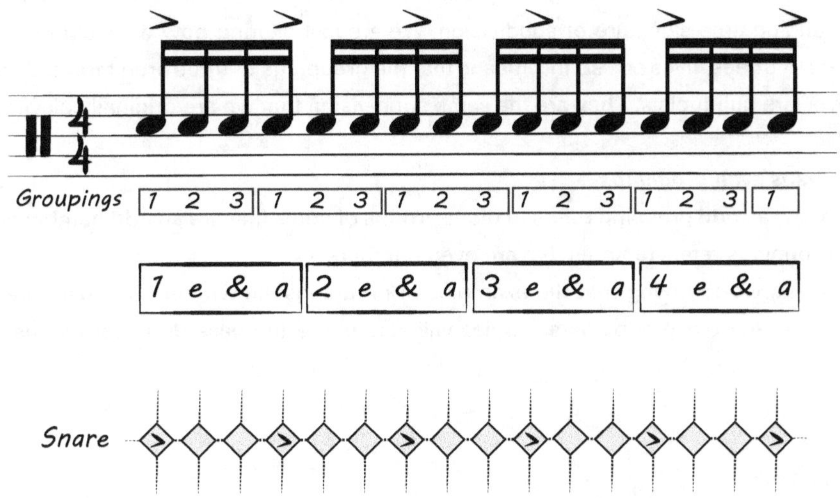

NB This phrase can naturally occur over the bar line; right now, however, we are going to feel it like this as a one-bar phrase (hence why when you loop it you end up with two accents in a row).

In the same way that we have divided our bar of 16th notes in 4/4 into groupings of three, we can also do the same but with groupings of five. However, instead of just putting one accent at the first of every five notes, we are going to also put an accent on the first and the third. Thus, our grouping of five is two and three put together. Check out the image below, look at where the accents fall and you'll see what I mean.

Groupings of five

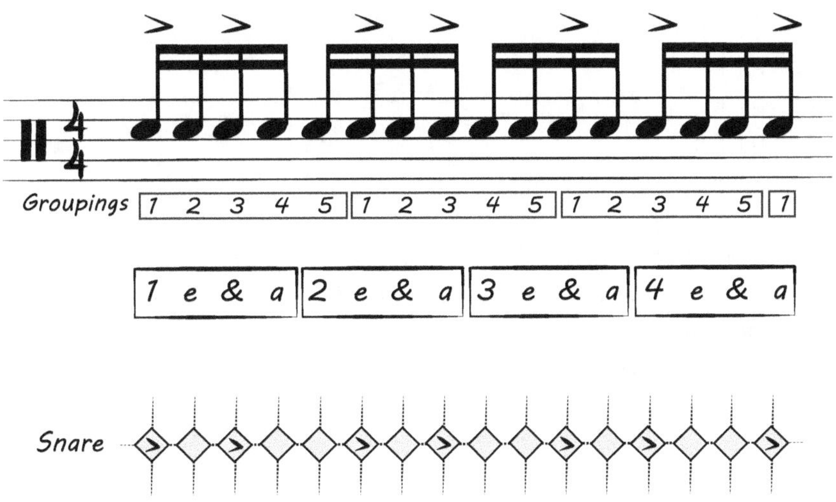

Drum Groove Creation

A note on groupings

Please be aware that when using odd groupings (or even ones), this does not mean that we are changing time signature or subdivision. We are just altering how we feel the notes as clusters of threes, fives etc. so this means that the groupings of three aren't triplets and that the fives are quintuplets. They are the same subdivision that we are originally playing in.

Odd versus even groupings

Put simply, an odd grouping refers to those groups of notes that are an odd number and the even groupings refers to notes that are even numbers.

It is also worth noting that the examples of groupings shown here are just on a snare drum. Some of our groupings examples will also utilise the bass drum to help us create interesting new grooves.

Key

R = Right hand
L = Left hand
R/L = Right or left (you choose)
K = bass drum (commonly referred to as the kick drum).

This key does is provided for producers as well as drummers. Choose a drum and assign a hand to it, i.e. right hand to hi-hats or left hand to snare drum. This will help you keep track of what you are creating. Then try altering the assigned drums up in steps 4 and 5.

Groupings

R = Right hand
L = Left hand
K = Kick drum

Groupings of 2 – R/L K
Groupings of 3 – R L K
Groupings of 4 – R L K K
Groupings of 5 – R L R L L

NB Once you are comfortable with my suggested stickings for groupings, try and experiment with your own. For the grouping of five, you could replace the last two 'L L' with a 'K K'. This will give you an alternative feel.

Also, R/L means you have the choice of right or left; it does not mean that both have to be played together (unless you want them to, of course).

A linear drumming method example one

1. Pick a time signature – ***4/4***
2. Pick a subdivision – ***16th notes***
3. Use groupings to make up a bar – ***3 5 3 5*** (=16)
4. Orchestrate and find a backbeat
5. Orchestrate around the drum set

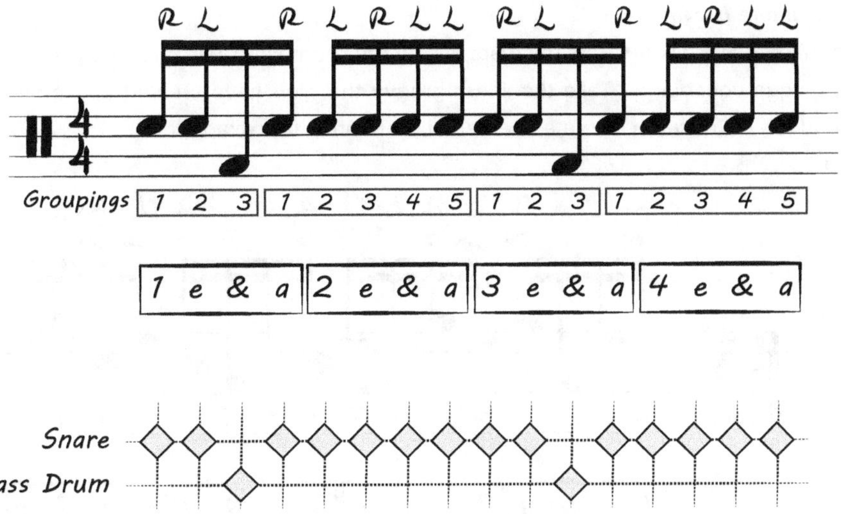

Musical mileage

I could quite easily dedicate a whole chapter to musical mileage. Here, however, it's worth discussing the importance of stretching an idea to get the most out of it. This process contributes to your ability to create grooves from staying on an idea before moving onto the next one. Experiment with different orchestrations of the linear groove and try placing the hands across different drums to create new and interesting drum fills. See the examples below for a few variations of this approach.

Groove 66: Linear 3 5 3 5

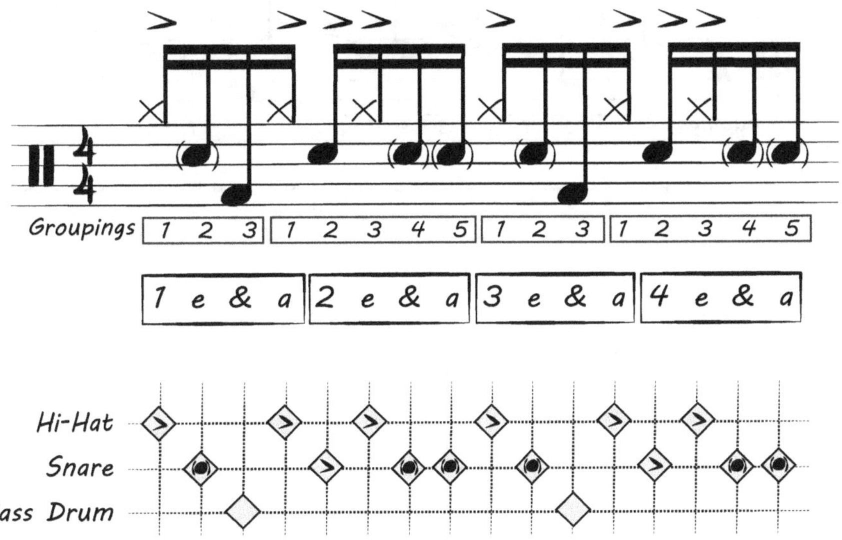

Orchestration and sound substitution

Explore these variations to help orchestrate this groove around the drum set and gain some more musical mileage out of it. This could be through taking the right or left hands and playing them on the floor tom. Another means of orchestrating the groove is to move the notes played on the hi-hats onto the ride cymbal.

Use as a one-bar fill

A lot of students ask me how to create fills on the drum set. Well, this a way to create new and unique vocabulary. Take the linear groove that you have created and orchestrate it around the drum set to create a one-bar fill. Here are a few examples.

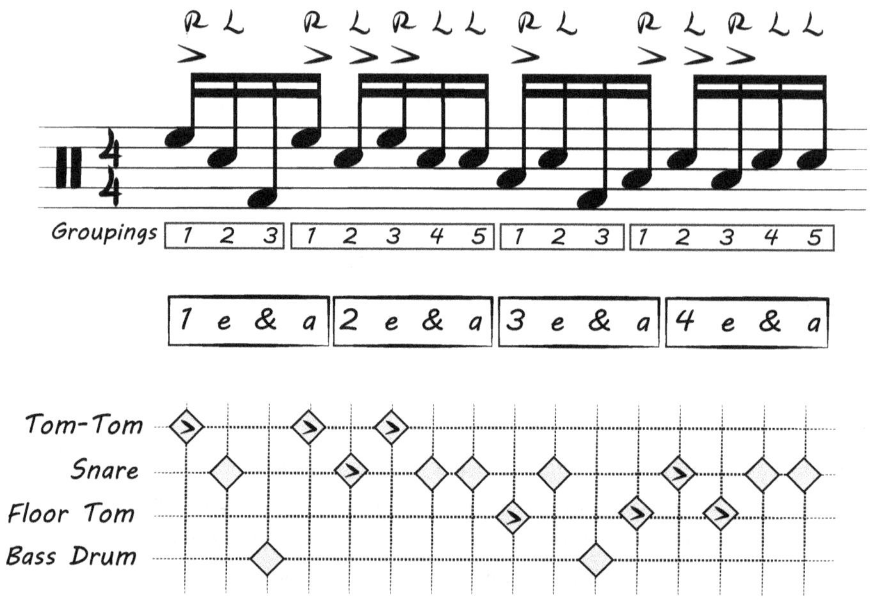

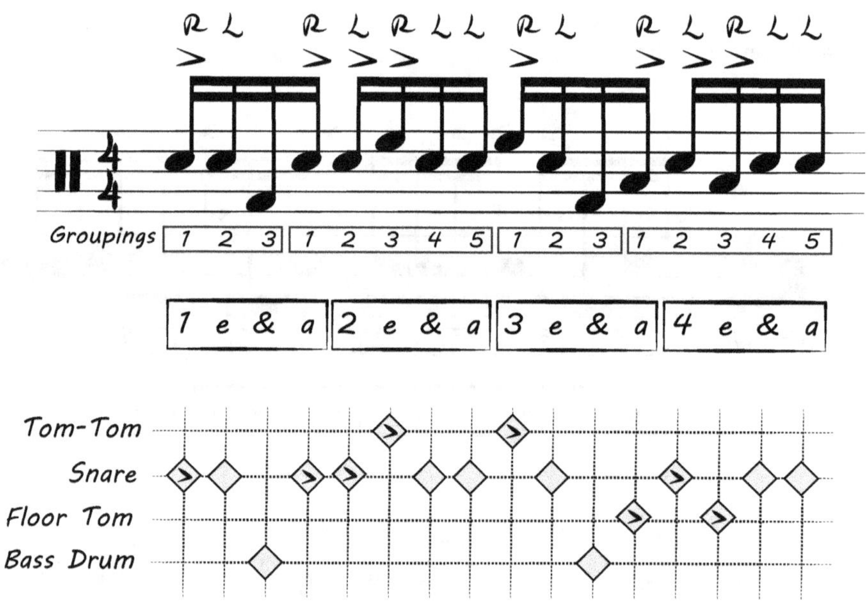

NB You will notice that I have put a suggested sticking. This is what works for me; if you are a left-handed drummer, you could reverse this.

Mixing up the groupings

Mix up the order of the groupings used in step 3 in the linear drumming method. This will help give you lots of variations from using the same numbers, in this case 3,5,3,5.

3 3 5 5
5 5 3 3
5 3 5 3

Of course, if you are playing different things for each '3' or '5', then that would also add to the amount of times you could re-arrange this. Equally, if you had a phrase with four different numbers, then you would get a lot more mileage out of it.

Groove creation: Challenge twelve

For this chapter's groove creation challenge, it really is a case of putting the linear drumming method into practice. This could take a long, long time if you do it properly and exhaust all of your ideas. So be patient, and experiment with different time signatures and subdivisions (especially the ones you aren't comfortable with).

Also, an important note: keep track of what you are creating and what you like. There is nothing worse than stumbling onto something brilliant and then losing it. Use the journal table below to help with this.

When	What	Tempo	Notes/Reflections

Chapter 13 Groove amalgamation

Overview

In the previous chapters of this book, we have explored and examined some of the key drum grooves from around the world in the 20th and 21st centuries, as well as how to start creating new things with them, as shown in the Groove Creation challenges at the end of each chapter.

Something that I often think gets overlooked is our ability to be merge pre-existing grooves to create new and previously unheard ones. This is what I call **Groove Amalgamation**. In this process, we are essentially taking some of the most historically and culturally relevant grooves and merging elements of them together. This will hopefully give us (and the world) previously unheard and unique drum grooves.

You could, in essence, do this with every single groove shown in this book. I have created the **Diamond Cutter Method** to help.

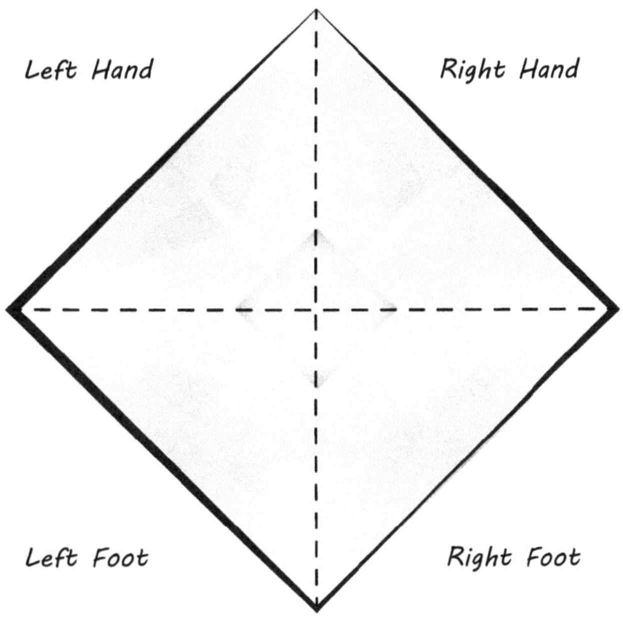

How the diamond cutter method works

This is our tool for how we use Groove Amalgamation. Each section of the Diamond represents a limb (this can be switched for left-handed drummers). You then will pick a different limb/individual instrument from different grooves in the book and put them together.

DOI: 10.4324/9781003585169-13

Drum Groove Creation

Producers/beatmakers, I know we don't necessarily think in terms of limbs but I do recommend it here just so we are able to separate elements of each groove. However, if you prefer you could think of each triangle of the Diamond as a separate drum/cymbal. It's up to you; whatever works best to help you amalgamate the previously shown grooves in this book.

Right Hand = Insert groove number/hi-hat option or ostinato
Left Hand = Insert groove number/hi-hat option or ostinato
Right Foot = Insert groove number/hi-hat option or ostinato
Left Foot = Insert groove number/hi-hat option or ostinato

NB Depending on the style or culture you are constructing, you may find that it affects your choices here. Something that can be quite challenging is what to do with the left foot (the hi-hat pedal). Equally, depending on what you are creating, you may choose to leave it dormant or replace it with a piece of percussion if you are creating this in a Digital Audio Workstation (DAW).

Below are some examples of how we can use the Diamond Cutter Method for Groove Amalgamation to discover and construct new drum grooves.

Groove amalgamation examples

Groove 67: Displaced Funk with Baiao

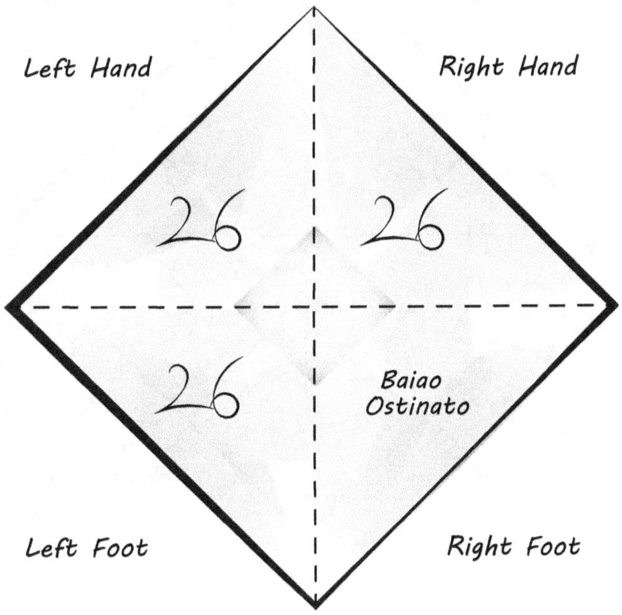

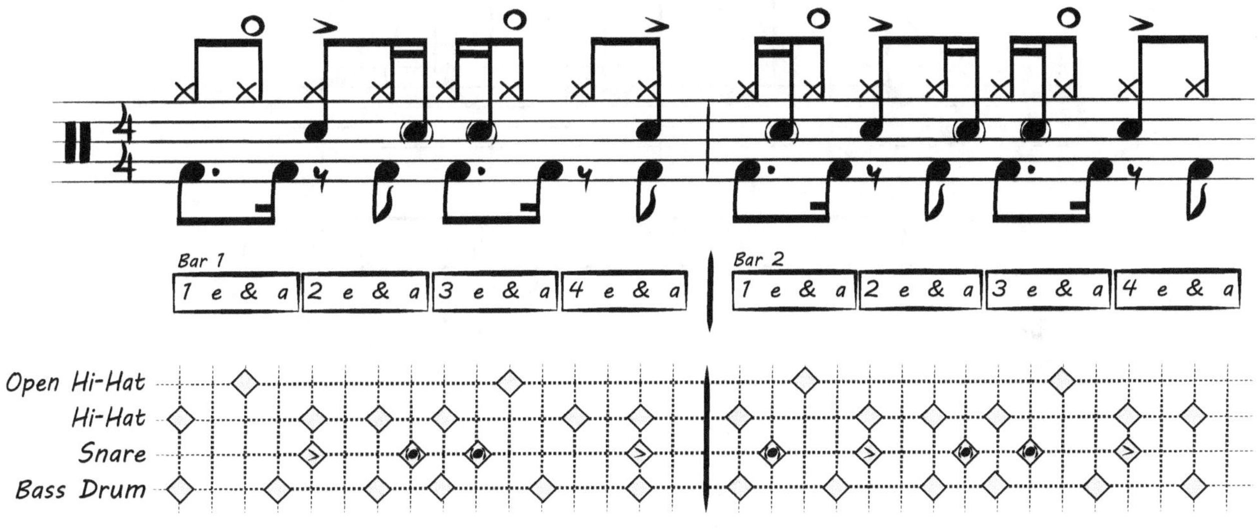

Here we have taken concepts from our James Brown-inspired Funk section and taken the Funky characteristic of the displaced backbeat and put it together with our Baiao ostinato.

Groove 68: Soul style stepped half time shuffle

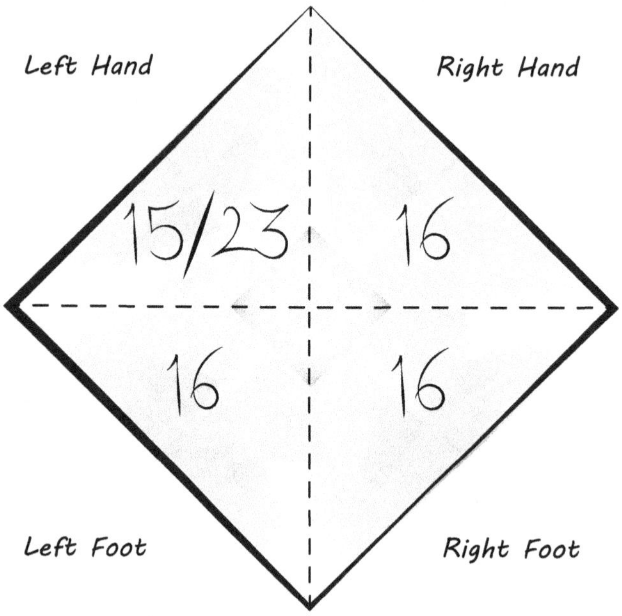

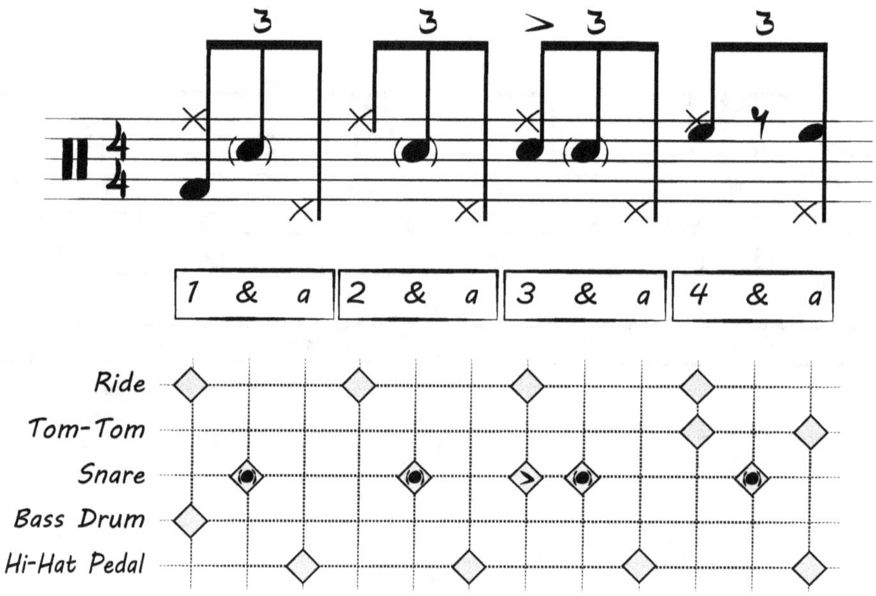

Here we are amalgamating our stepped shuffle pattern but playing it on the ride cymbal (instead of the hi-hat, as shown in the original stepped shuffle example). The backbeat is on beat three, just like the half time shuffle, and we have added the last two notes of our Al Jackson Jr-inspired soul groove example one we looked at from our Soul chapter.

Groove 69: Songo & Baiao

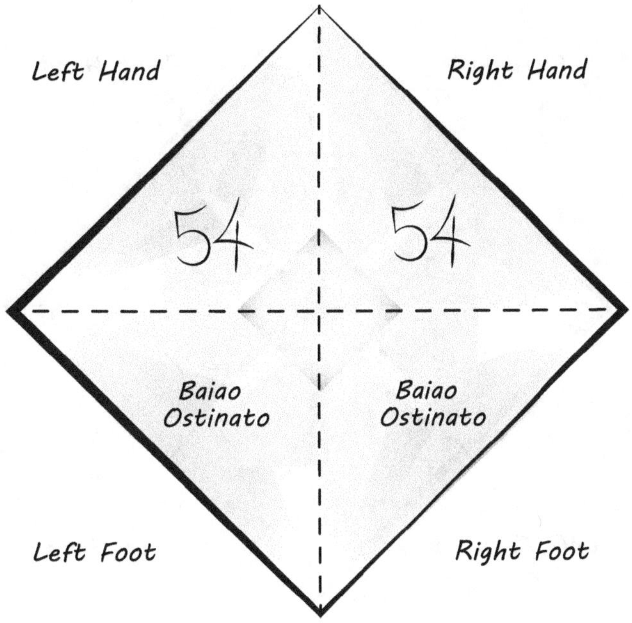

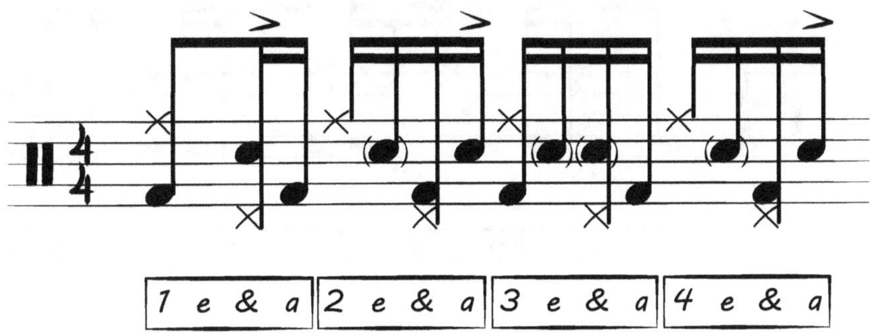

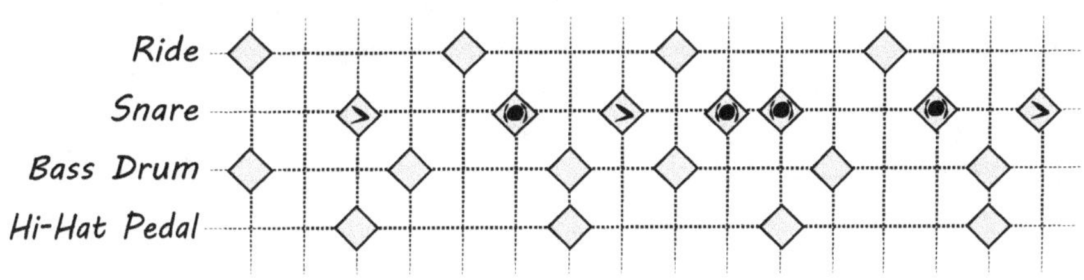

In our second example of groove amalgamation, we see what happened when we utilise elements from different cultural rhythms. The snare and ride are playing a classic Afro-Cuban Songo pattern, and the feet are playing the Baiao foot ostinato.

Groove 70: The Samba Guaguanco

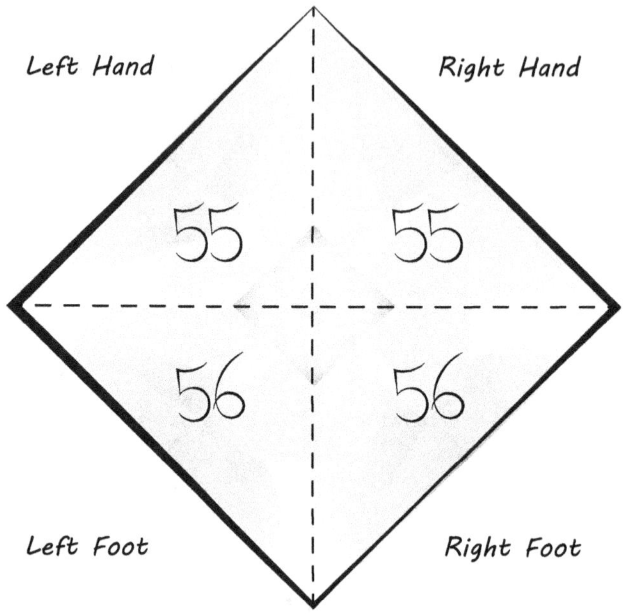

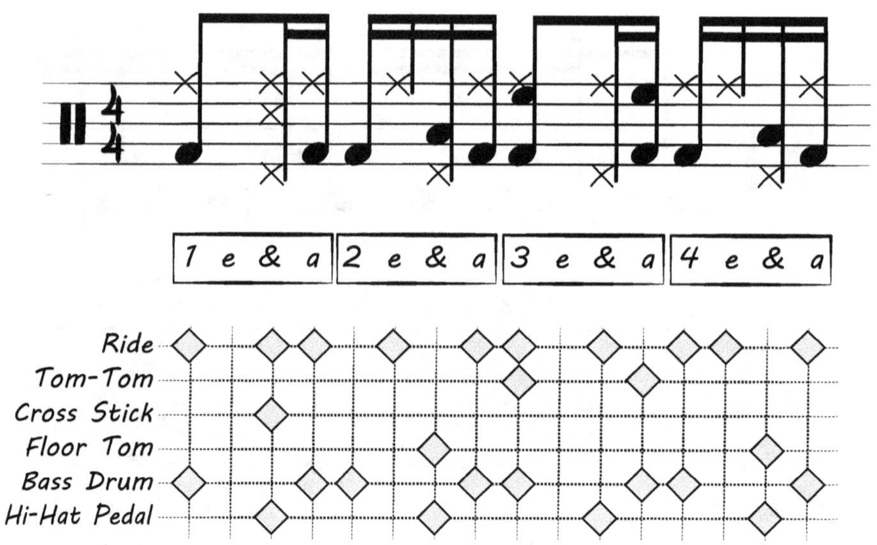

This is quite a fun example that I like to experiment with. Here we are taking influences from Africa, Cuba and Brazil. The *Samba Guaguanco* consists of the bass drum and hi-hat pedal from our traditional Samba pattern and our snare and ride playing the cascara pattern and also the guaguanco tom tom melody. There is quite a lot being combined in this big melting pot.

Musical mileage in groove amalgamation

The next few examples explore how we can get more mileage out of our Groove Amalgamations. Here I am taking the concept of the Mozambique and seeing how it could be used in different groove contexts.

Groove 71: The tumbao Mozambique

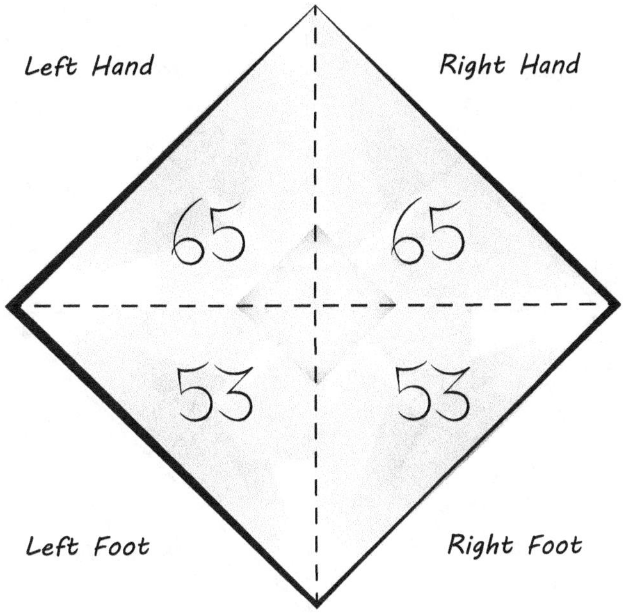

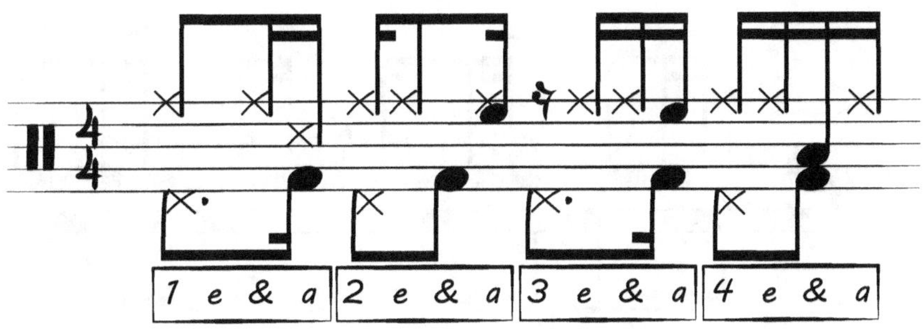

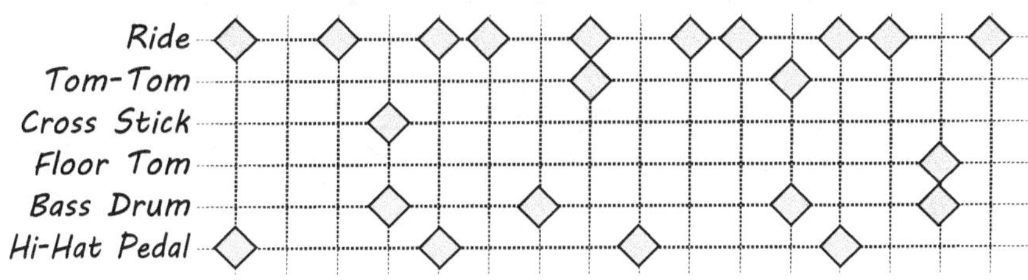

These are quite fun grooves, where we take the Mozambique pattern and add a few different flavours. These are just a few examples of how you can take a concept from a single groove and transform it into something new and exciting.

Groove 72: Tumbao, Mozambique, ghost notes and backbeat

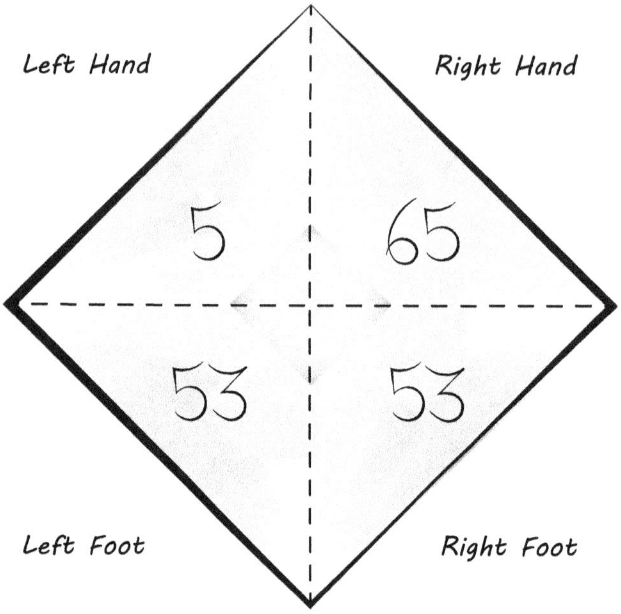

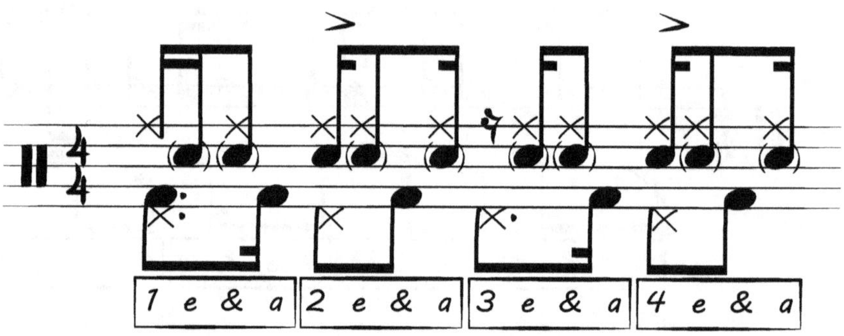

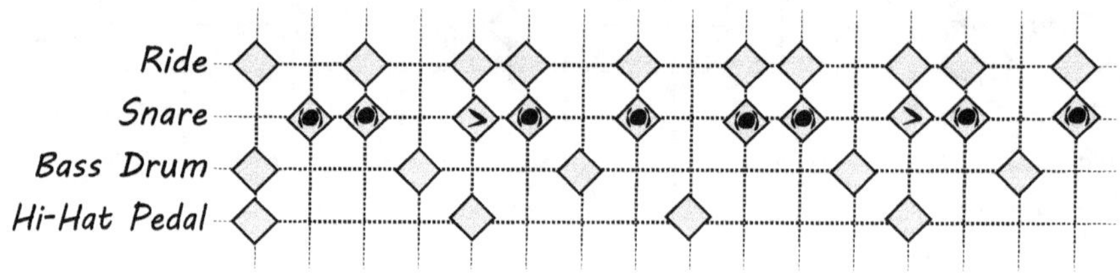

In this one we take the Mozambique hand pattern, keep the tumbao foot pattern from our Songo groove, but also add some Funk characteristics. Please note the ghost note pattern. This is quite a busy one, so be careful with dynamics between the ghost notes, backbeat and ride cymbal.

Groove 73: The displaced Funky Mozambique

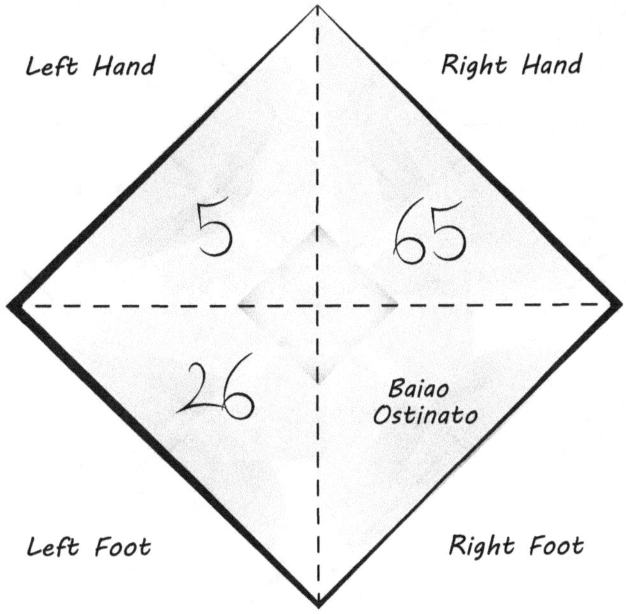

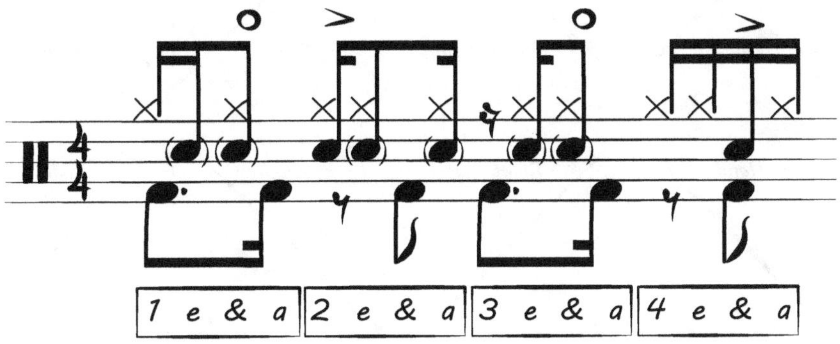

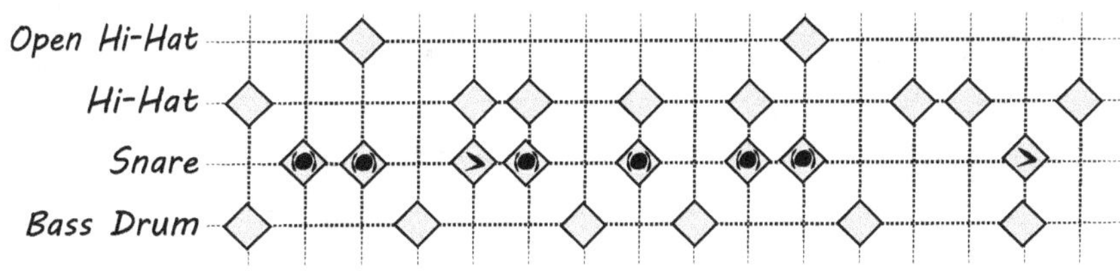

This is similar to the example above, except here we have taken some of the other characteristics from the Funk chapter. This time the Mozambique pattern is played on the hi-hats with openings on '1+' and '3+' as well as the displaced backbeat on '4+'. The right foot is playing the baiao bass drum pattern.

Groove creation: challenge thirteen

Your section

This is where you can explore groove amalgamation yourself. Fill in the sections with numbers from previous grooves in this book, use the Diamond Cutter Method and keep a record of what you have created.

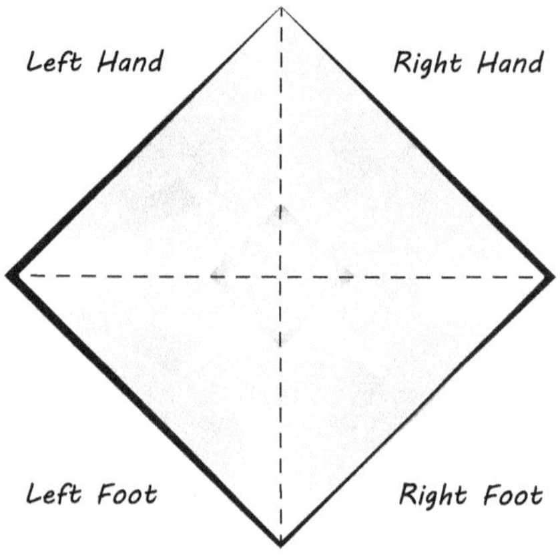
Left Hand
Right Hand
Left Foot
Right Foot

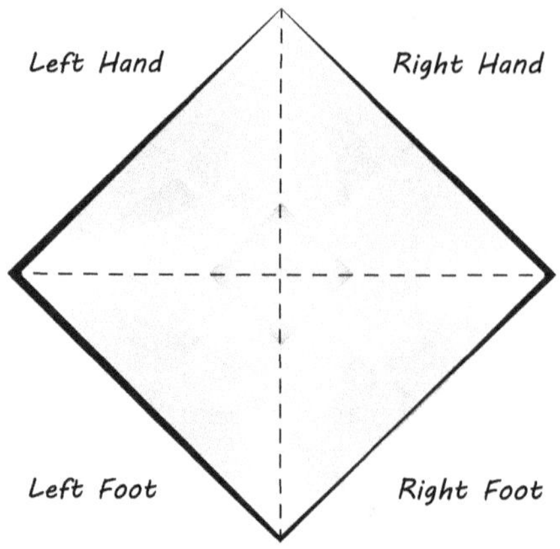
Left Hand
Right Hand
Left Foot
Right Foot

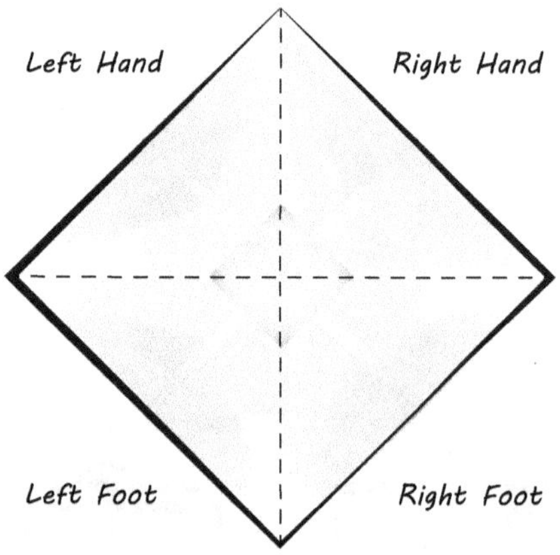
Left Hand
Right Hand
Left Foot
Right Foot

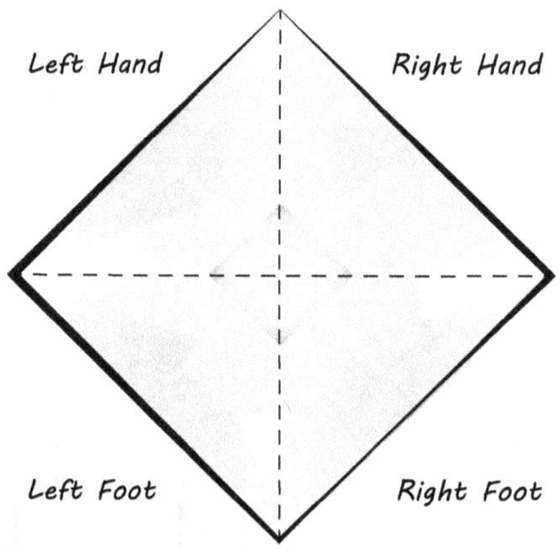
Left Hand
Right Hand
Left Foot
Right Foot

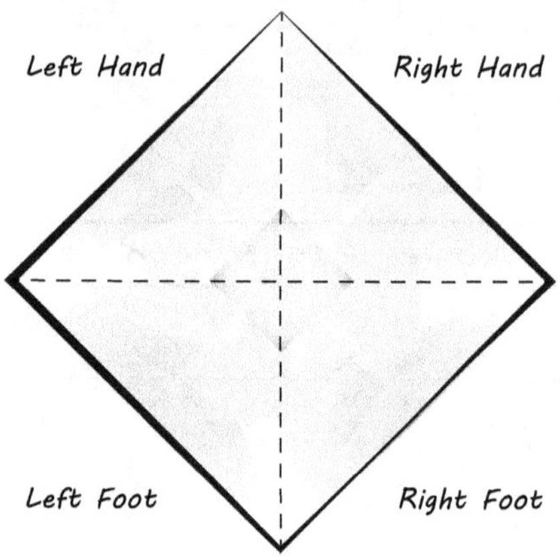
Left Hand
Right Hand
Left Foot
Right Foot

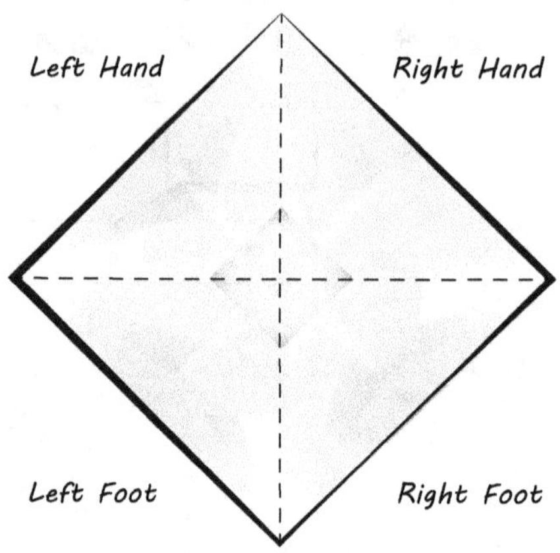
Left Hand
Right Hand
Left Foot
Right Foot

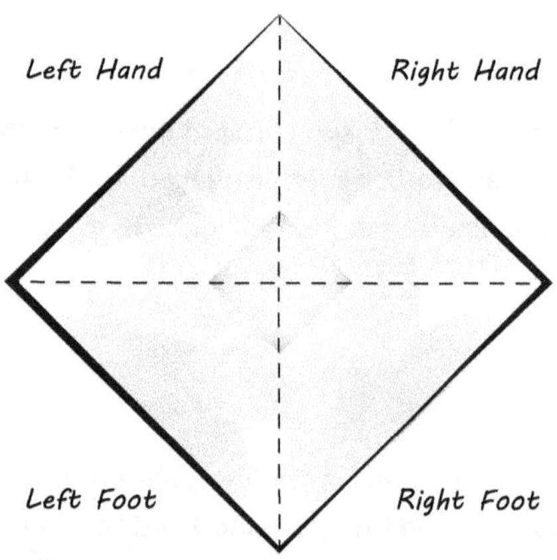
Left Hand
Right Hand
Left Foot
Right Foot

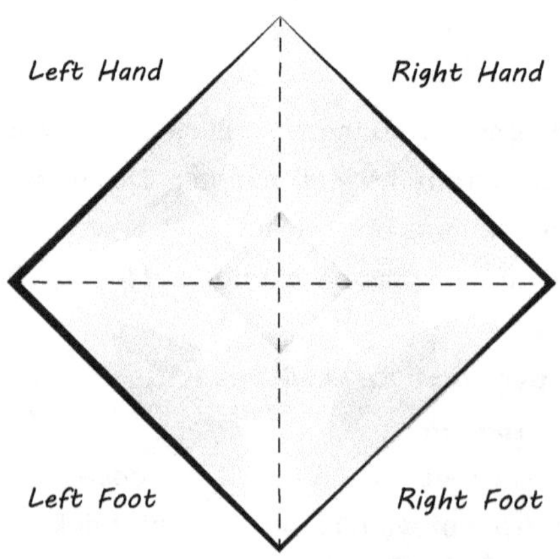
Left Hand
Right Hand
Left Foot
Right Foot

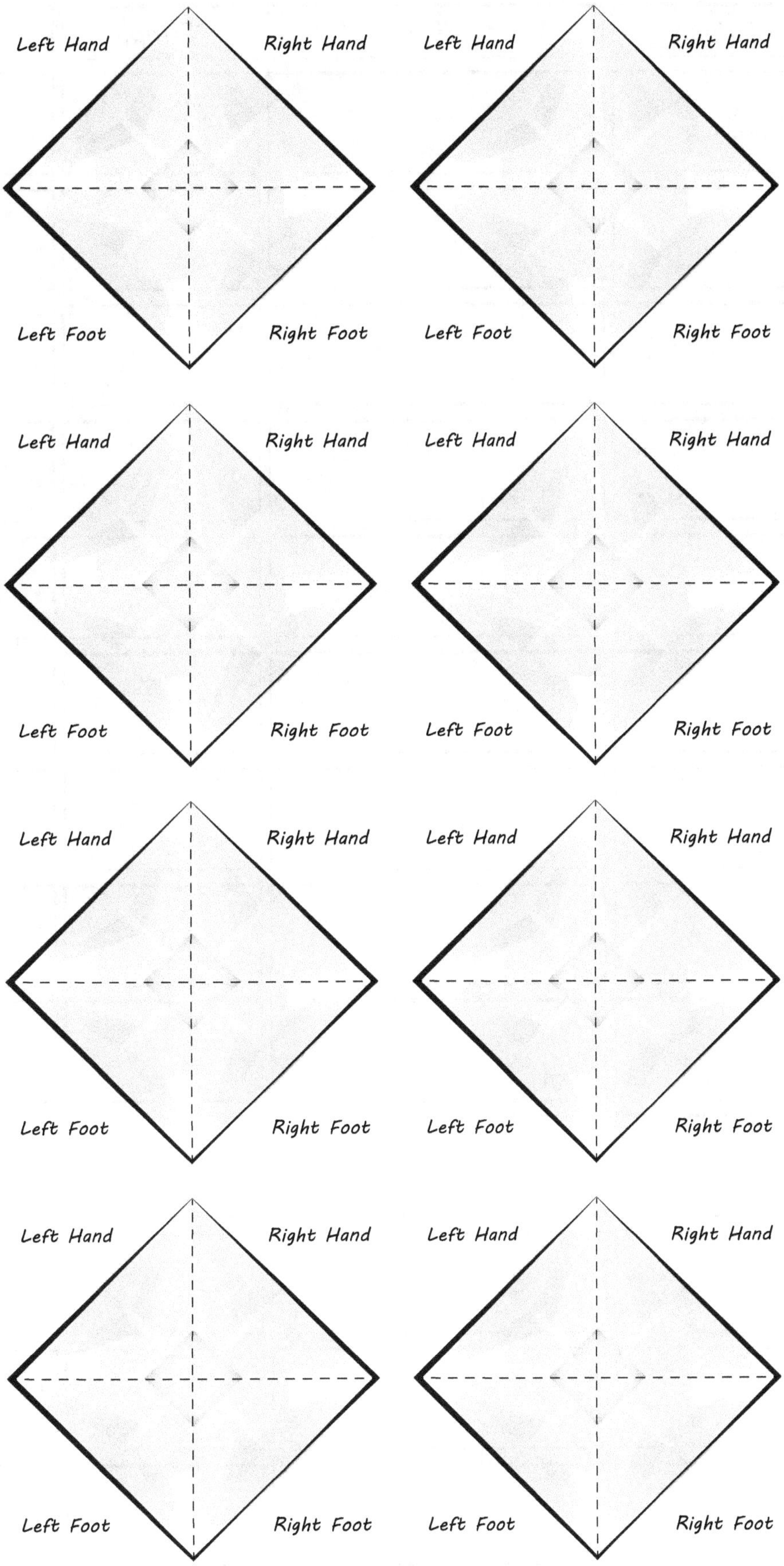
Left Hand
Right Hand
Left Foot
Right Foot
Left Hand
Right Hand
Left Foot
Right Foot
Left Hand
Right Hand
Left Foot
Right Foot
Left Hand
Right Hand
Left Foot
Right Foot
Left Hand
Right Hand
Left Foot
Right Foot
Left Hand
Right Hand
Left Foot
Right Foot
Left Hand
Right Hand
Left Foot
Right Foot
Left Hand
Right Hand
Left Foot
Right Foot

When	What	Tempo	Notes/Reflections

Chapter 14 Drum groove creation

Overview

So far in the book we have examined grooves from different drummers, artists, and genres and considered how they fit together. It is now time for us to examine the method that will allow us to create our own grooves. Below is a guide and set of instructions on how to begin creating your own drum grooves.

Groove creation elements

Step 1 The timekeeper pattern
Step 2 The snare pattern
Step 3 The bass drum pattern
Step 4 The hi-hat pedal

Step 1: Selecting the timekeeper

This is what is commonly played on the hi-hats, ride, crash, floor tom or perhaps a piece of percussion such as a shaker.

Option 1 – Straight 8th notes

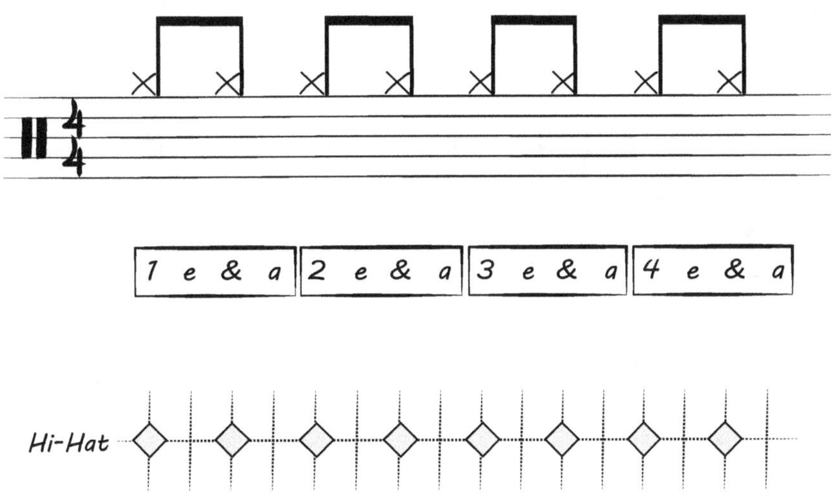

DOI:10.4324/9781003585169-14

Drum groove creation

Option 2 – Digging in

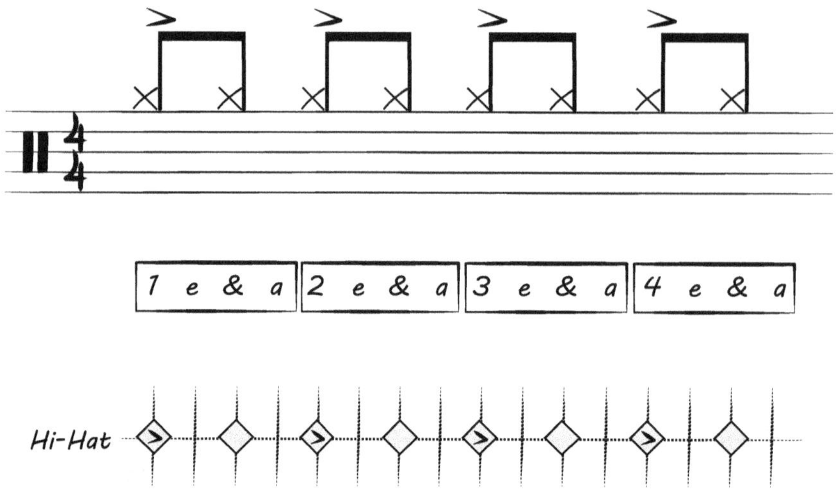

Option 3 – Upbeat feel

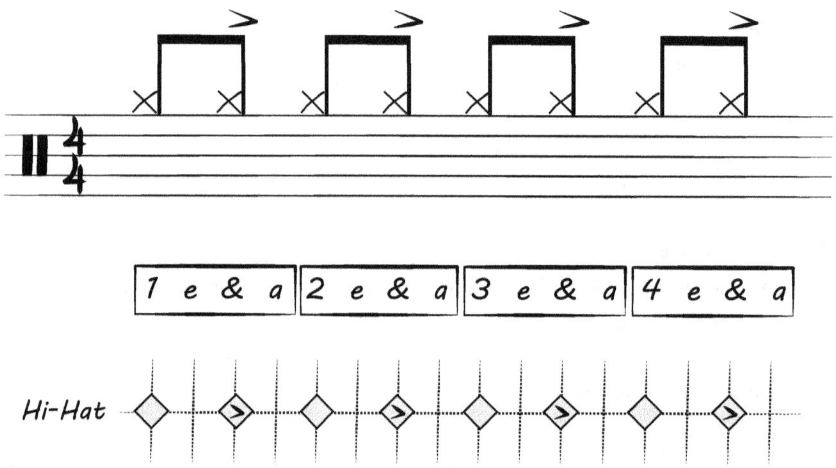

Option 4 – 16th notes

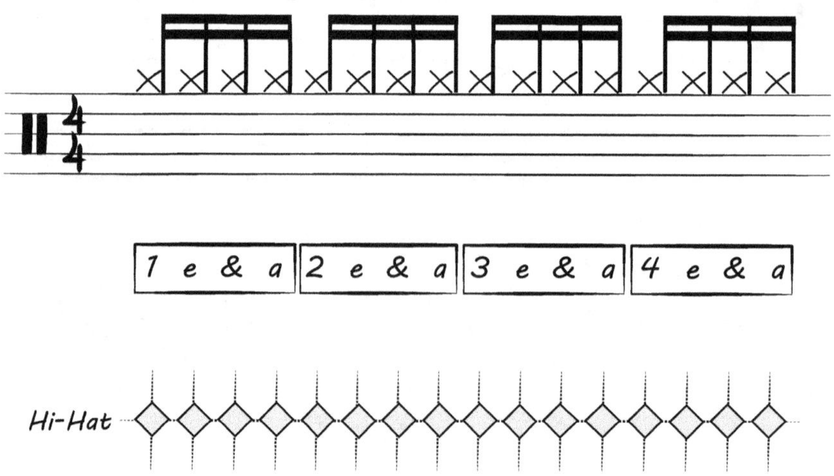

Option 5 – Broken 16th note pattern A (1 + ah)

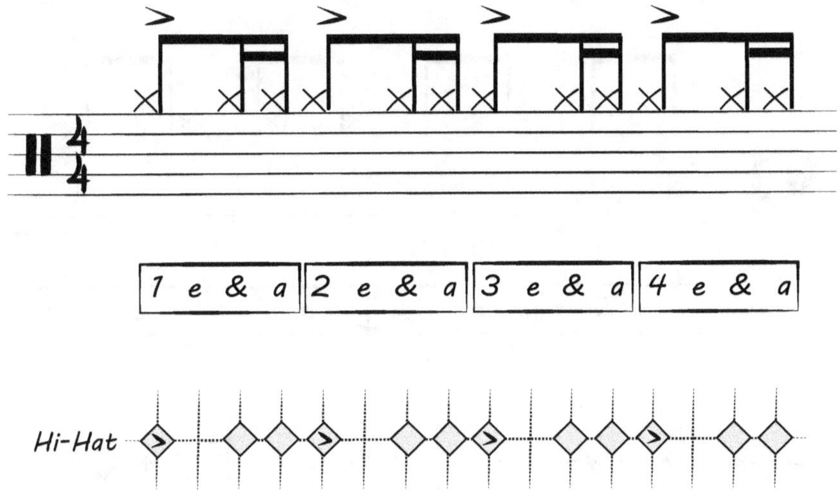

Option 6 – Broken 16th note pattern B (1 e +)

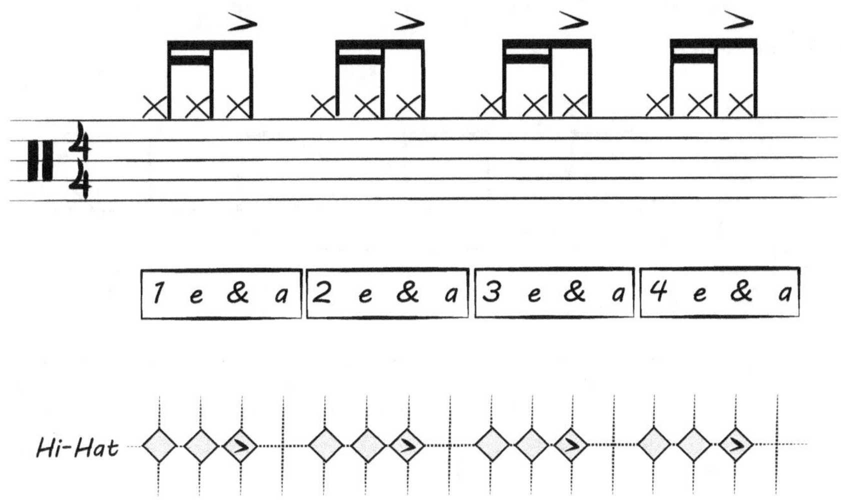

Option 7 – Quarter notes

This may seem like the simplest option, but sometimes the simplest route is the most effective.

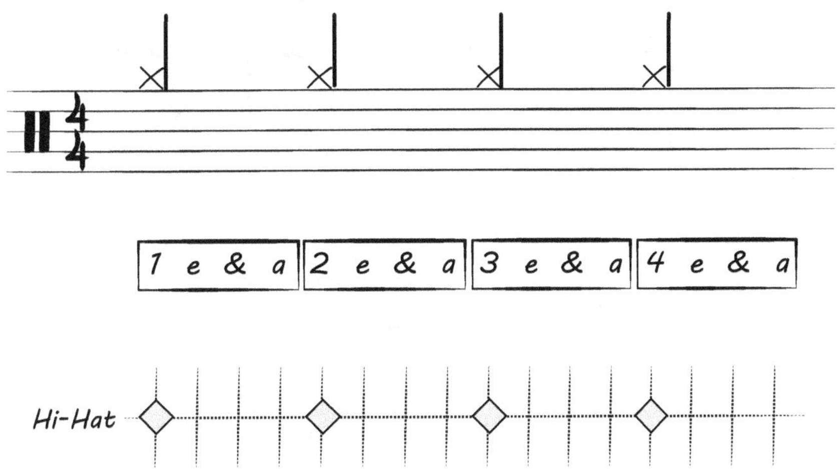

Drum groove creation

Option 8 – Moving the accents

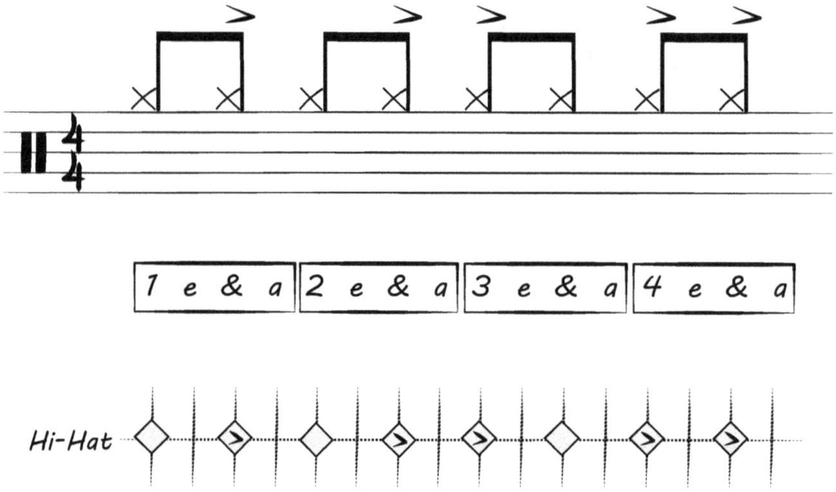

As shown in our Funk chapter

Funk hi-hat option A

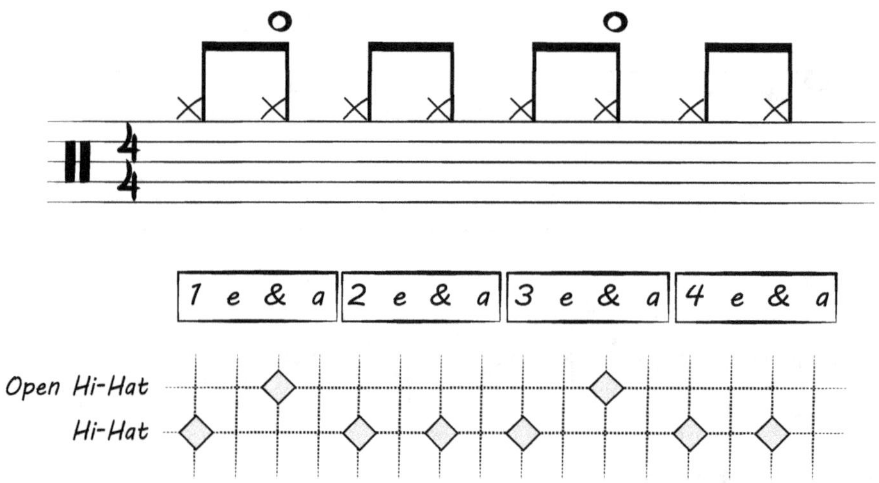

Funk hi-hat option B

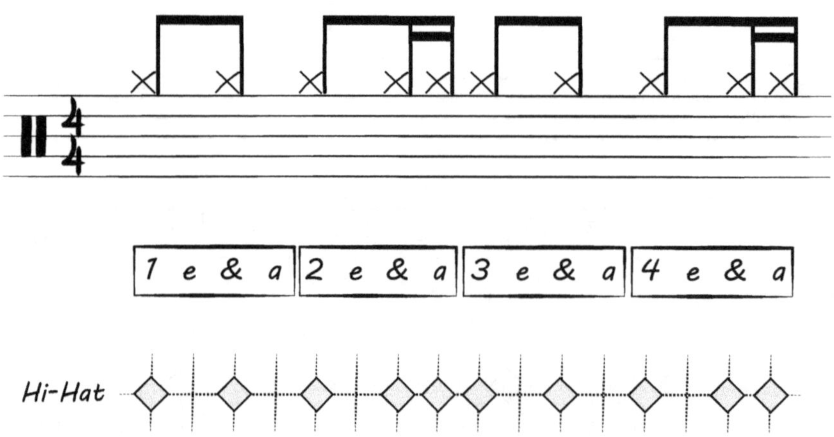

Funk hi-hat option C

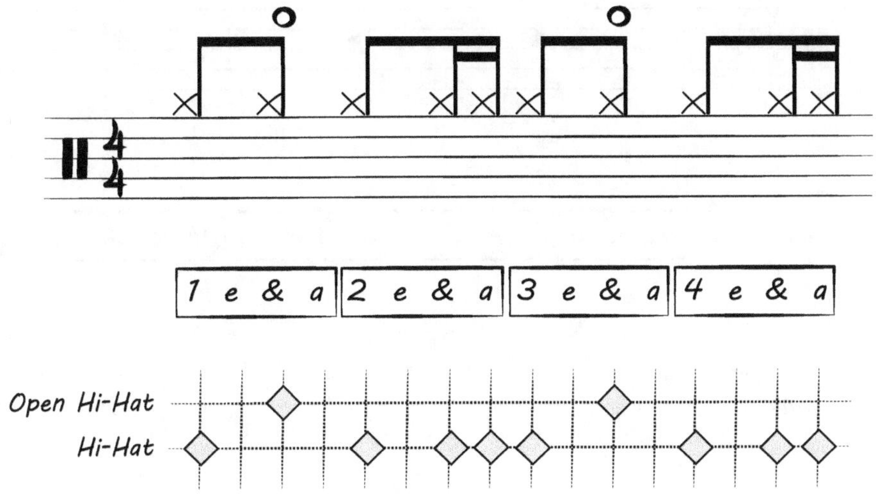

Some of these have been explored in our grooves already in this book, so there should be some familiarity here. Others are new, so take them as they come.

Challenge: There is always the option of crafting some of your own timekeeper patterns if you desire. Start with these ones and go from there.

Step 2: The snare pattern

After you have decide the timekeeper, consider what pattern you want to be played on the snare drum.

Option 1 – The backbeats (snare on 2 and 4)

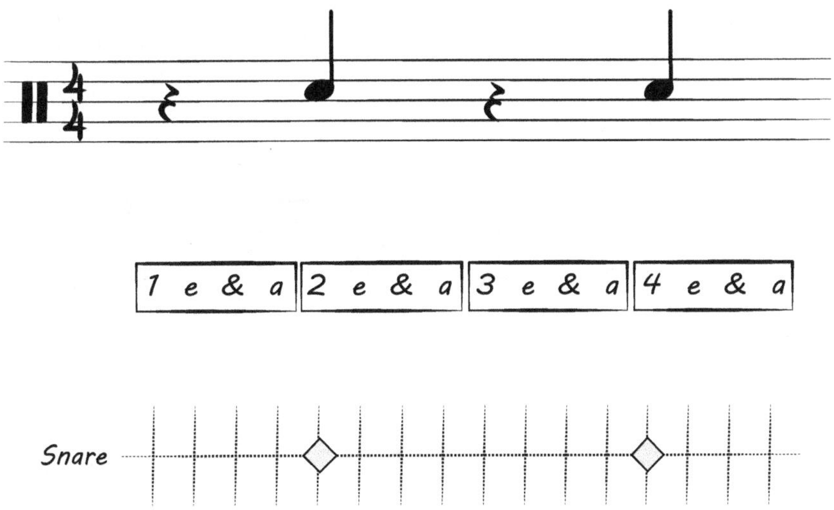

Drum groove creation

Option 2 – Displaced backbeats (snare on 2 and 4+)

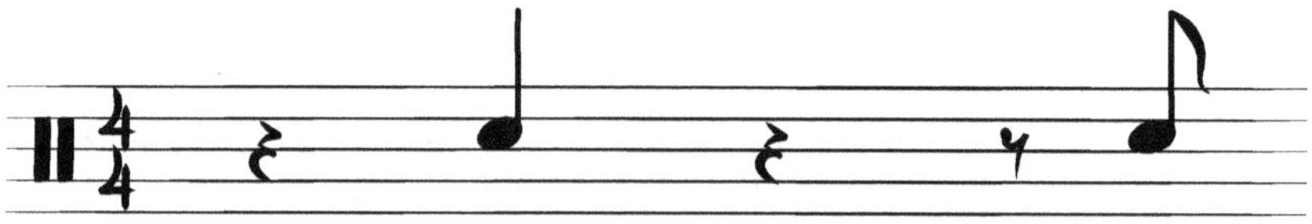

1 e & a	2 e & a	3 e & a	4 e & a

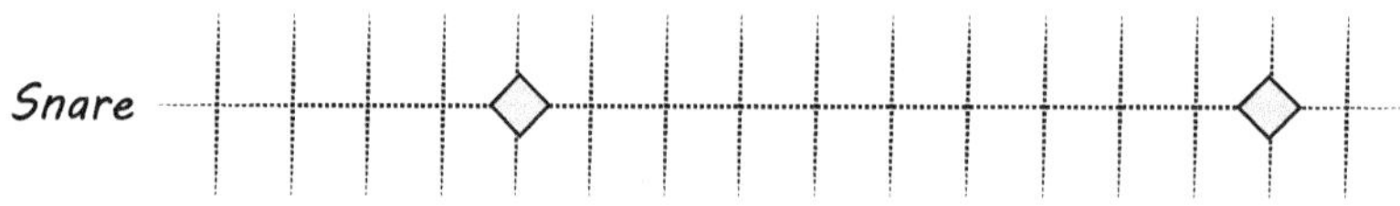

Option 3 – Ghost note pattern A

1 e & a	2 e & a	3 e & a	4 e & a

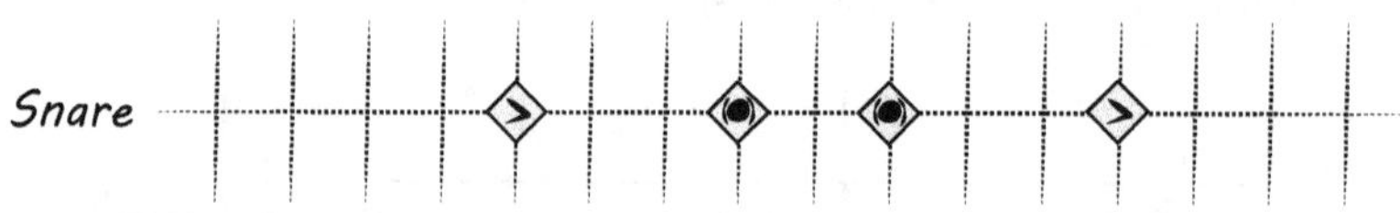

Option 4 – Ghost pattern B

1 e & a	2 e & a	3 e & a	4 e & a

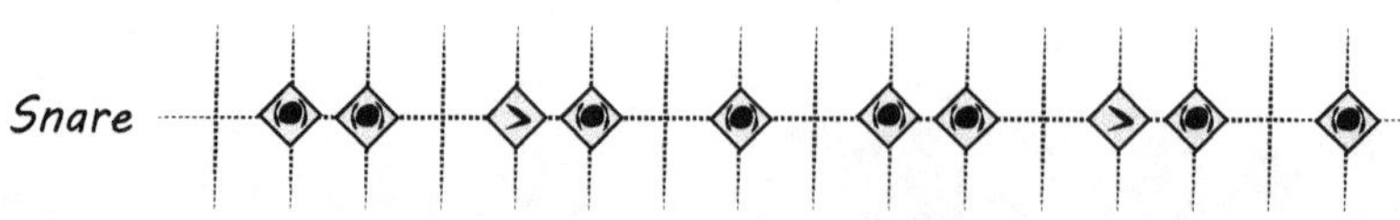

Step 3: The bass drum pattern

Option 1: The next step in our groove construction is adding the bass drum. For this, I would recommend revisiting the 16th note placements that was discussed at the end of Chapter 1. This will help you create your own patterns and bass drum melodies.

Option 2: Equally, another approach, in a similar way to groove amalgamation, is to take some of the bass drum patterns from grooves previously shown within this book.

Option 3: In terms of a music creation approach, we must also consider how our grooves interlock with other instruments in the context of a track. In our third option, it is more a shift in mindset. Here we are constructing the bass drum pattern based on what is happening in perhaps a bass guitar on a track. This could be a track we are creating ourselves in a DAW or perhaps in a live band scenario.

Step 4: Hi-hat pedal options

Some of these have been touched on throughout the book, but I wanted to compile them in one place. Here are some placements for us to play the hi-hat pedal. These can be used to help amalgamate grooves and fill in the bottom left of the Diamond Cutter Method discussed in our groove amalgamation chapter.

Hi-hat pedal optional 1 – 8th notes

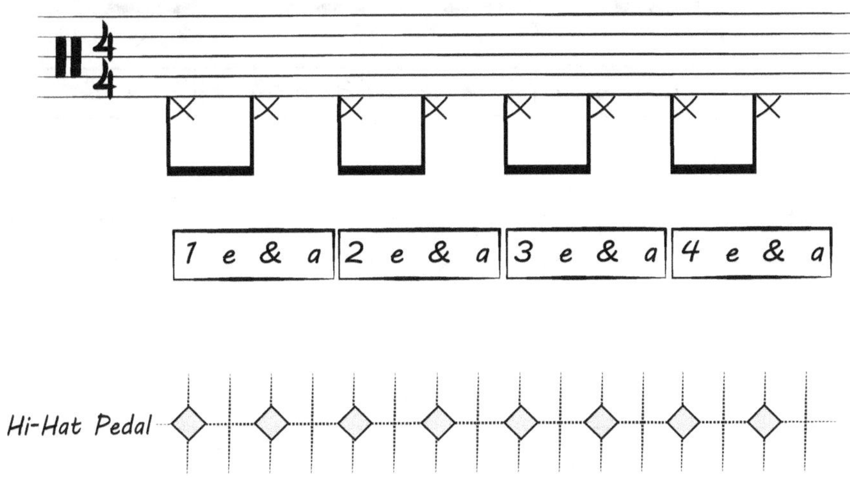

Hi-hat pedal optional 2 – Quarter notes

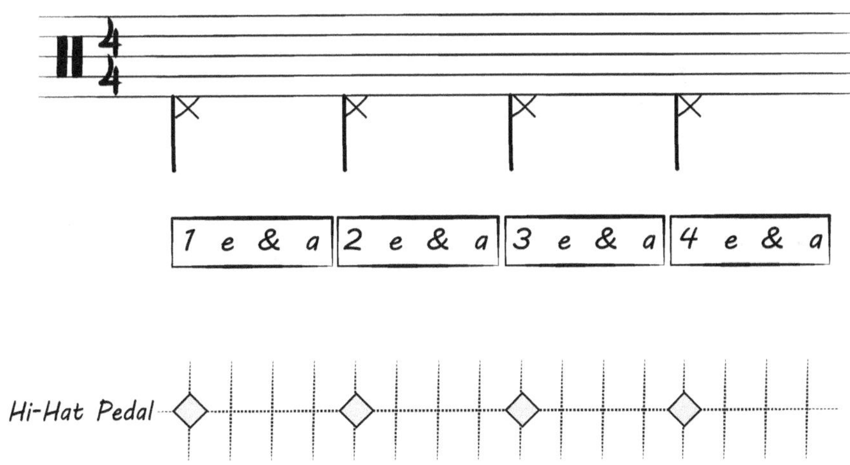

Drum groove creation

Hi-hat pedal optional 3 – Upbeats

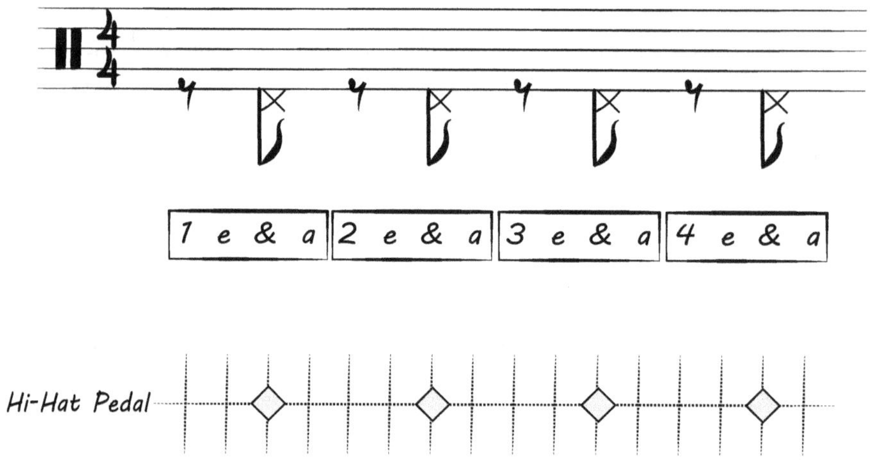

Hi-hat pedal optional 4 – Splash/close

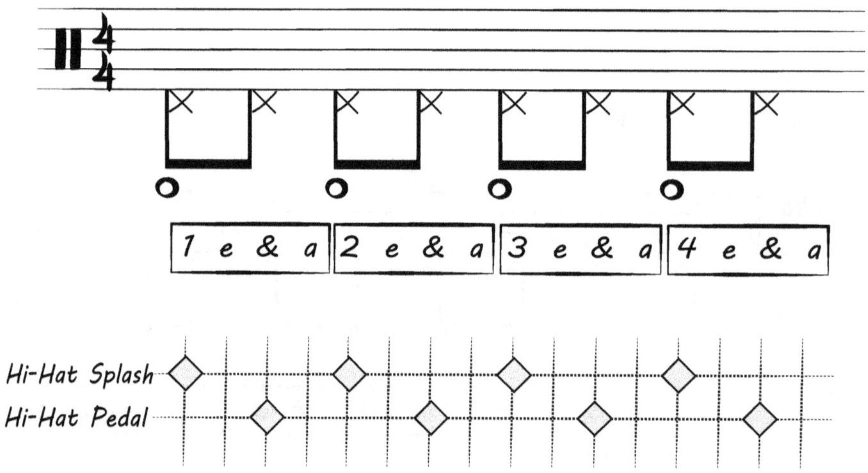

Hi-hat pedal optional 5 – Two and four

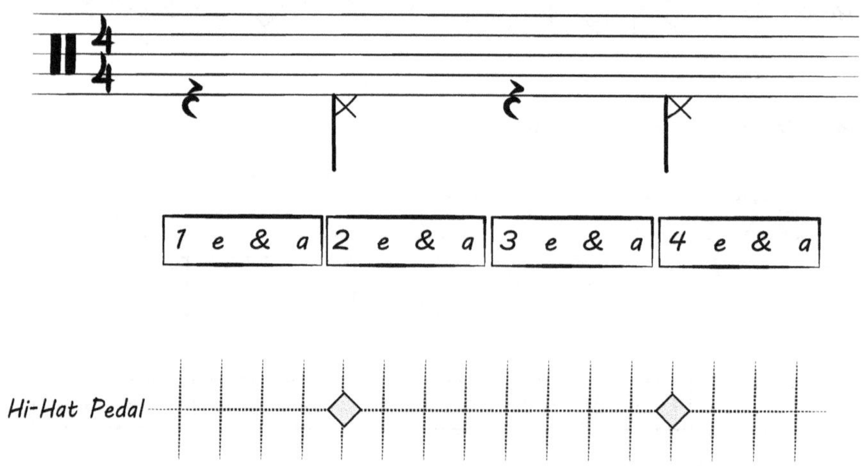

The Diamond Cutter Method in drum groove creation

In a similar way that we used our Diamond Cutter Method in groove amalgamation we can apply the same principle of thinking for drum groove creation. For drum groove creation, we will use each triangle from the diamond for each of our four steps (timekeeper, snare, bass drum and hi-hat pedal).

Timekeeper = right hand
Snare drum = left hand
Bass drum = right foot
Hi-hat pedal = left foot

Some examples to get you started...

Here are some examples of putting **groove creation** into practice. Start off by trying these ones out and then move onto creating your own using the method described.

Let's start off with a relatively simple one before we go too crazy with all of these options presented to us.

Drum groove creation

Groove 74: Cowbell and displaced backbeat

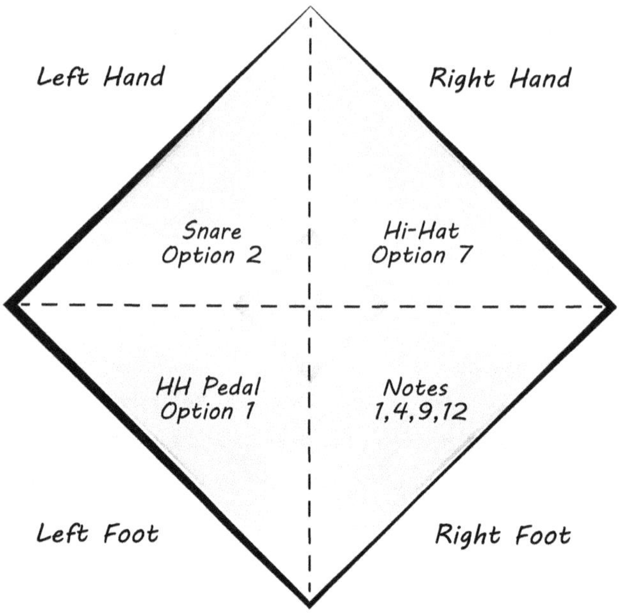

Step 1: Timekeeper

Cowbell on quarter notes

Step 2: The snare drum

Backbeat on beat 2 and on 4+

Step 3: The bass drum

In terms of our 16th note grid, 1, 4, 9 and 12. Refer back to Groove creation: Challenge one if you need a reminder of this concept.

Step 4: Hi-hat pedal

8th notes

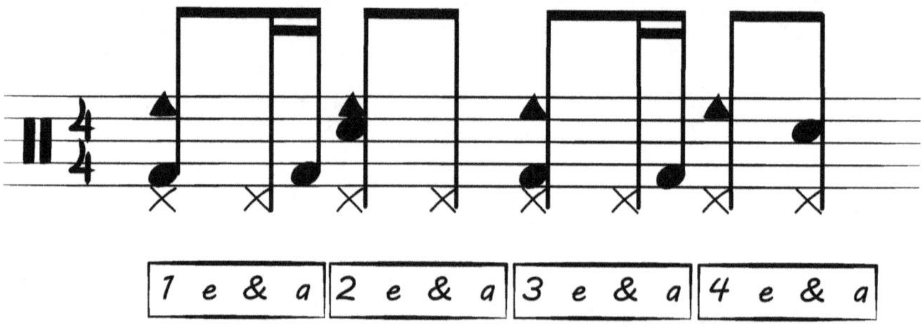

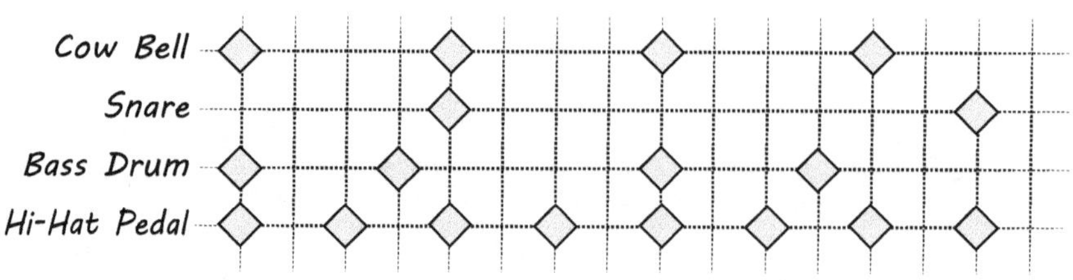

Groove 75: Ghost notes with broken 16th note rhythms

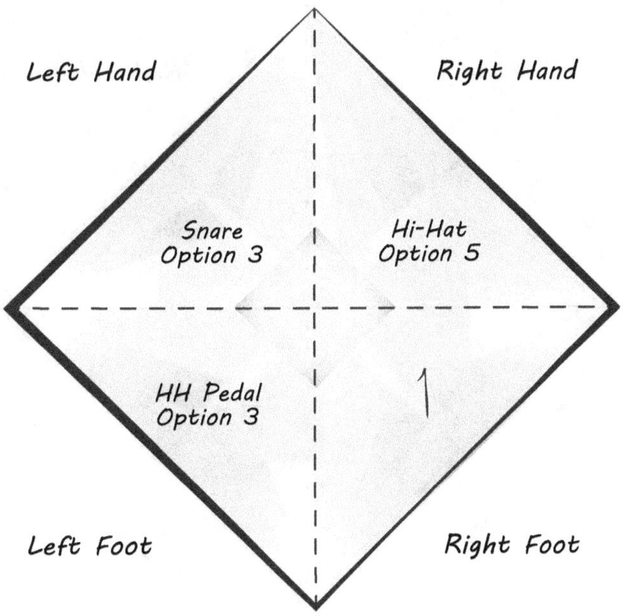

Step 1: Timekeeper

I have continued to use option 5 here, playing it on the ride cymbal and then adding the downbeats on the bell of the ride cymbal.

Step 2: The snare drum

I then added ghost pattern A, which is snare drum option 3.

Step 3: The bass drum

I kept the bass drum pattern simple and took the bass drum pattern from Groove 1, 'The Money Beat'.

Step 4: Hi-hat pedal

Here I elected upbeat on the Hi-hat pedal to give an uplifting and danceable feel. This is hi-hat pedal option 3.

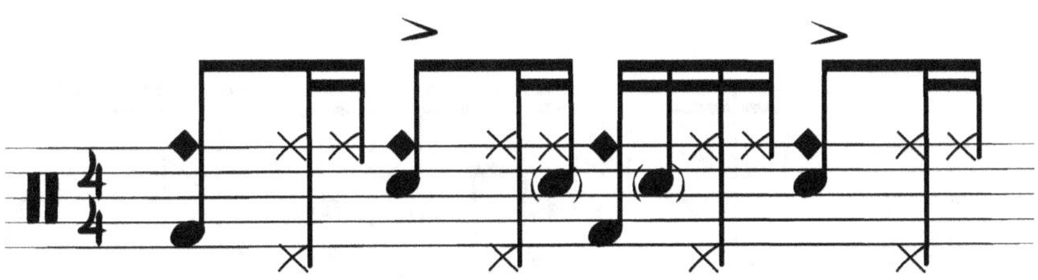

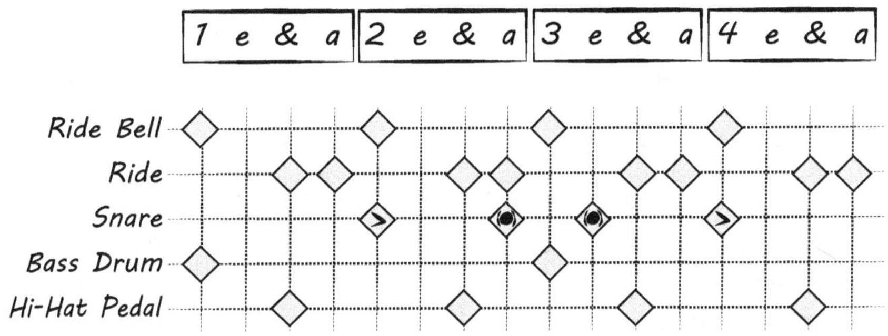

Drum groove creation

Groove 76: Ghost notes B with broken 16th note rhythms variation

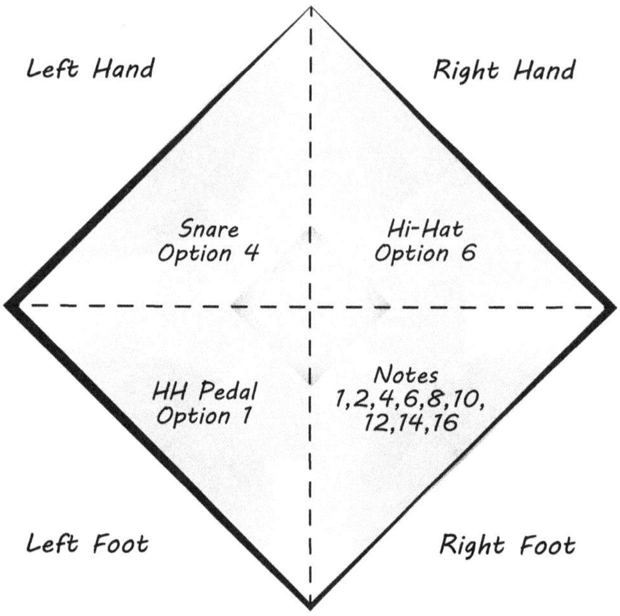

Step 1: Timekeeper

I have taken hi-hat option 6, put it on the ride cymbal, but every 'and' (every third note in other words), I have placed on the bell of the cymbal.

Step 2: The snare drum

I then added ghost pattern B (snare drum option 4) for some variation to the previous example.

Step 3: The bass drum

I used our 16th placements to create a melody (as described in Chapter 1). I have kept the downbeat on beat one, but then added all the upbeat 16th notes. In other words, every other note in our 16th note options.

Step 4: Hi-hat pedal

Here I elected to give this groove some drive, so I put all the eighth notes on the hi-hat pedal.

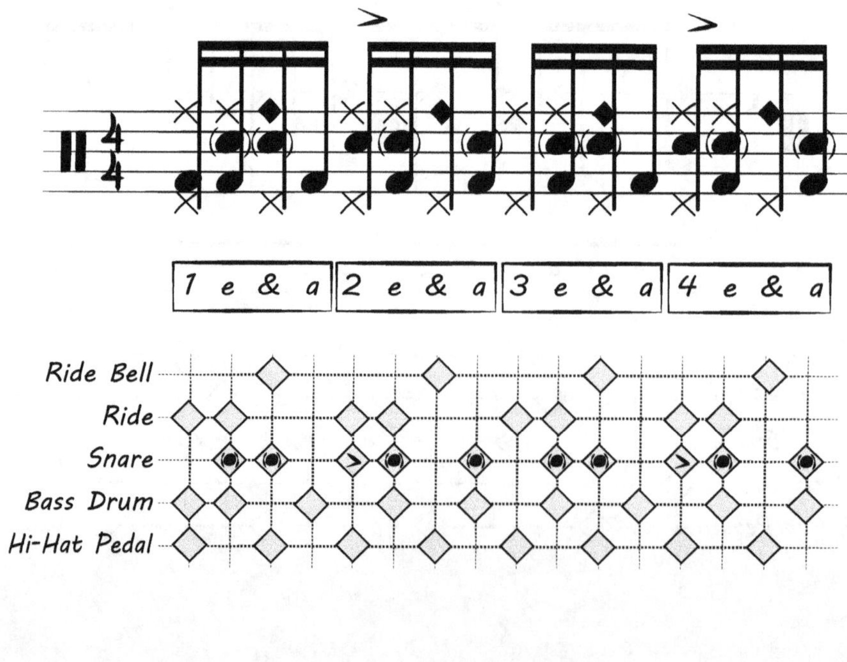

The infinite expansion on groove creation

Shown above are literally just a few of thousands of examples you could choose from. I cannot emphasise enough how many grooves you are able to create.

If you think of all the timekeeper options, each of these combined with a different snare drum pattern. Not to then mention the plethora of bass drum melodies you can create with our 16th placements grid.

Then, on top of all of these, every different combination of snare and timekeeper options with these hundreds (if not thousands) of bass drum variations can have additional patterns added with the hi-hat pedal.

Taking groove creation to the next level

So far we have been using the tools from this chapter. However, we can alter and expand on these ideas to add some more personality to our grooves. Check out the example below and explore how you could use these tools with some of the grooves that you create.

Drum groove creation

Groove 77: Moving the accents

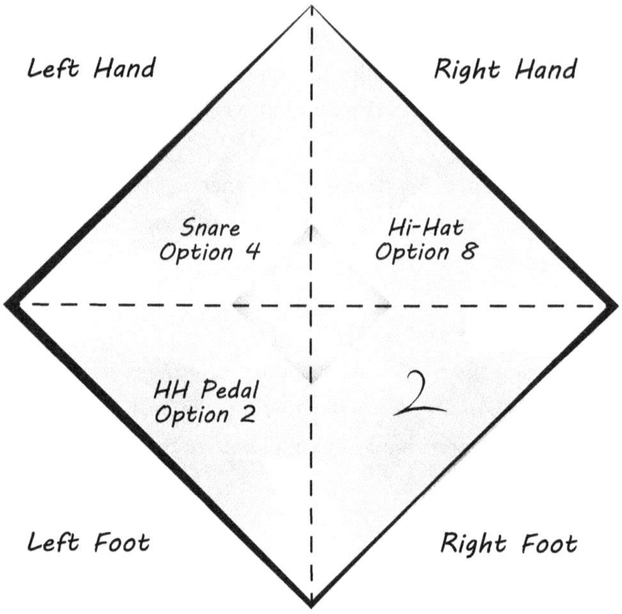

Something we haven't yet explored is adding and shifting accents; in this example it occurs on the ride cymbal. Note the shift of accents from timekeeper option 8 moved to the ride and bell of ride. This is a Stewart Copeland-inspired ride pattern with some ghost notes brewing under the surface.

Step 1: Timekeeper

Option 8 applied to the ride cymbal with the use of the bell.

Step 2: The snare drum

Snare drum option 4, ghost note pattern B.

Step 3: The bass drum

The ride and snare are quite busy in this groove, so I opted for the bass drum pattern from Groove 2.

Step 4: Hi-hat pedal

Hi-hat pedal option 2, steady quarter notes.

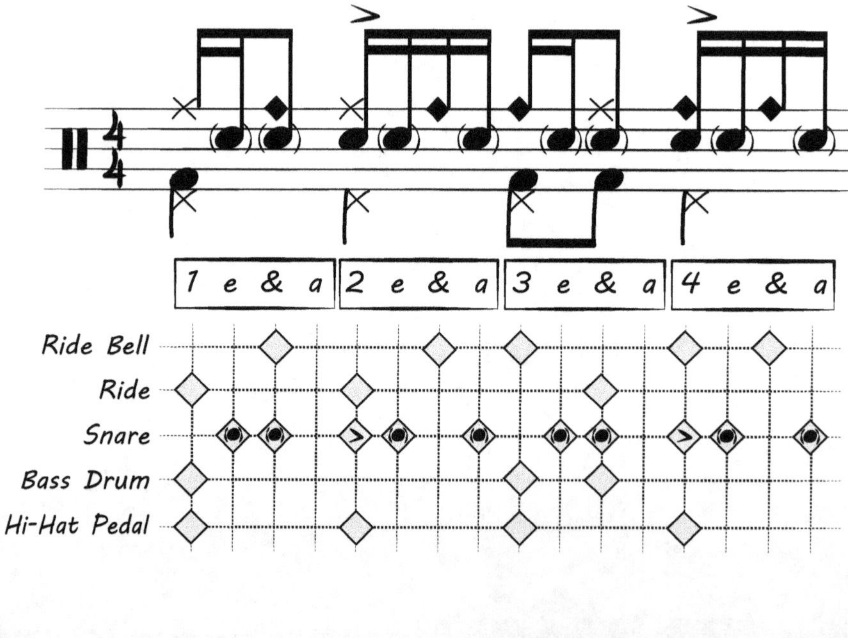

Groove 78: Moving the accents with ghost notes

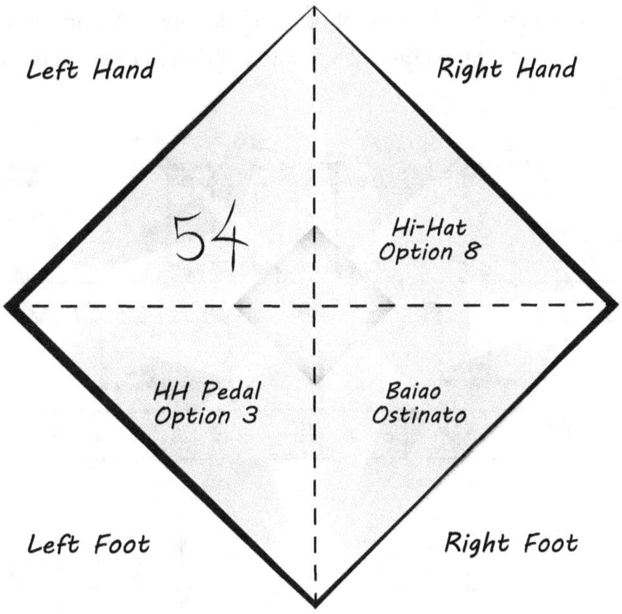

Step 1: Timekeeper

Here we are using option 8 again.

Step 2: The snare drum

This is the Songo snare drum pattern with a slight variation to the accents.

Step 3: The bass drum

This uses the Baiao bass drum pattern.

Step 4: Hi-hat pedal

Here we are using upbeats on the hi-hat pedal, which coincidentally works as part of the baiao foot pattern.

Drum groove creation

Here I have taken a Songo-inspired snare pattern and altered the accents slightly. Just to make things more interesting, I have taken the previous ride pattern from the last example to create something new. I have also thrown in the Baiao ostinato for good measure. This is quite a nice one to explore some different hi-hat pedal options with the Baiao. Having the hi-hat pedal play constant eighth notes can create a different effect in our grooves.

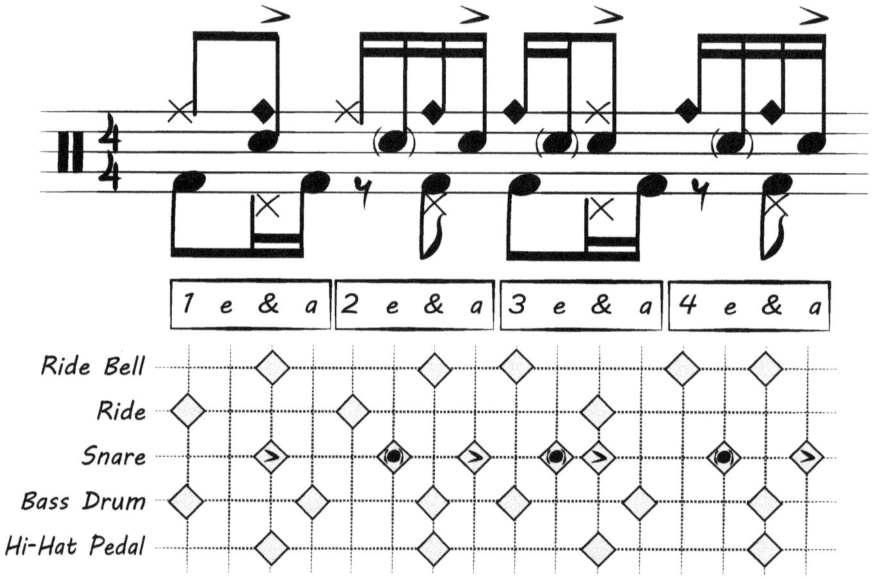

Groove creation: Challenge fourteen

Your section

In a similar way to groove amalgamation, as shown through this chapter, we can also use the Diamond Cutter Method to create new grooves. Your challenge is to use this chapter to create your own grooves.

Drum groove creation

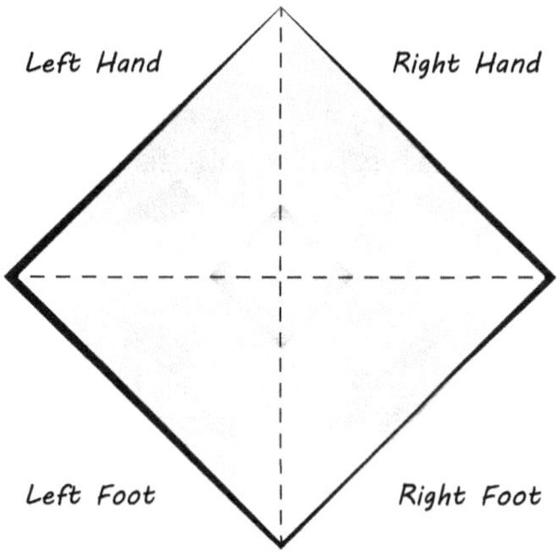

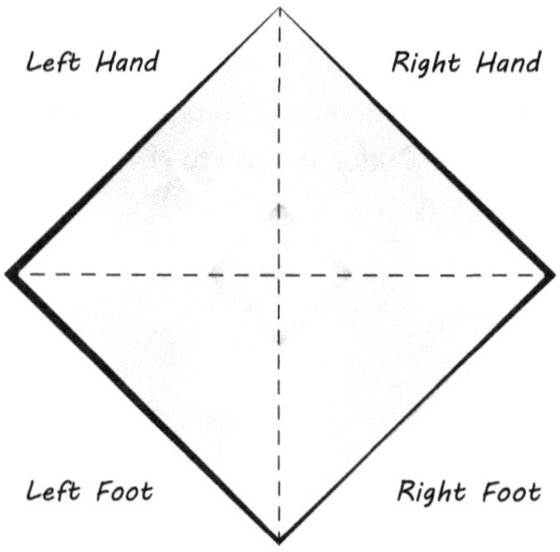

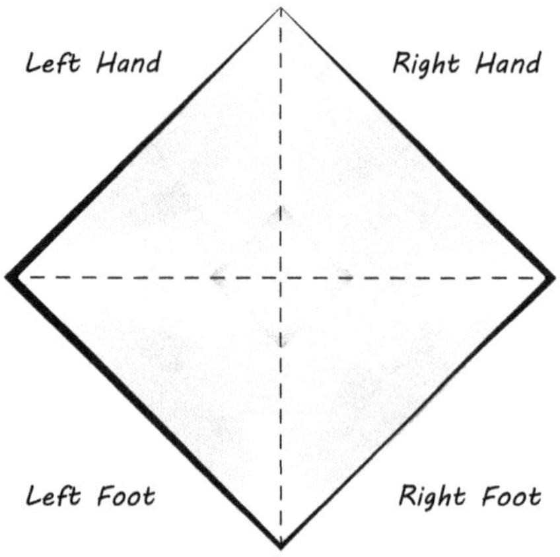

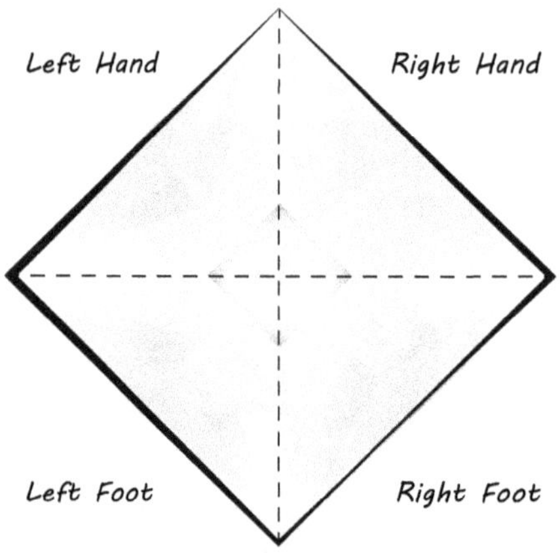

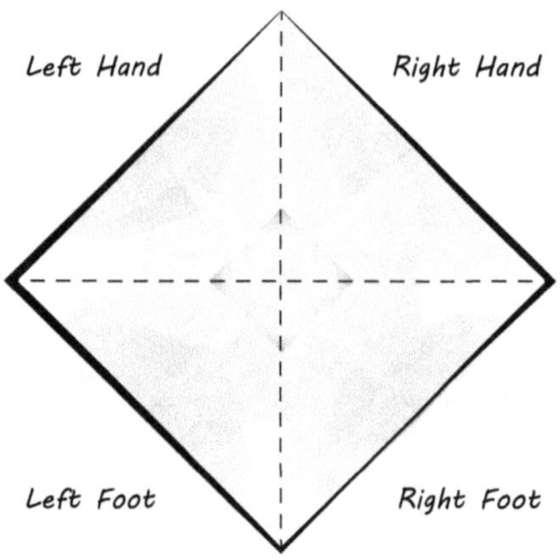

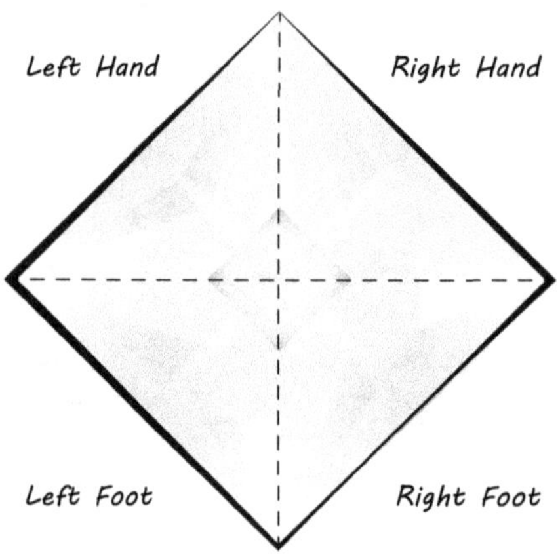

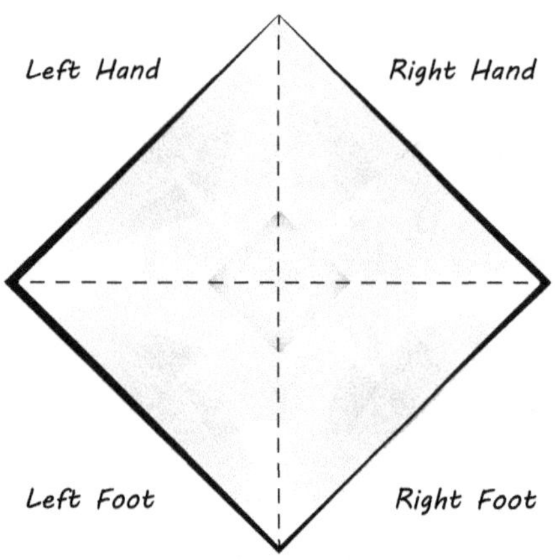

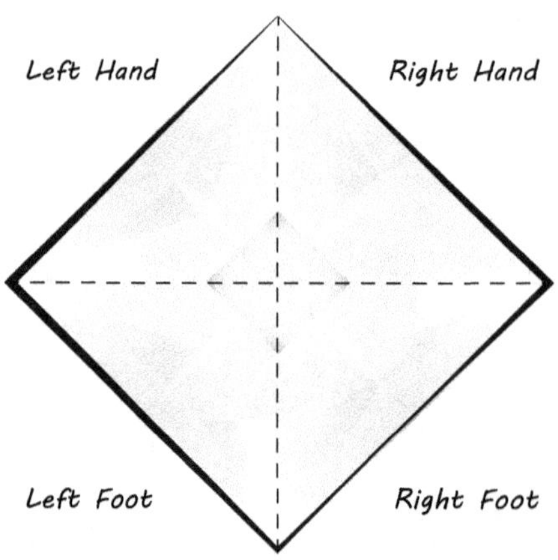

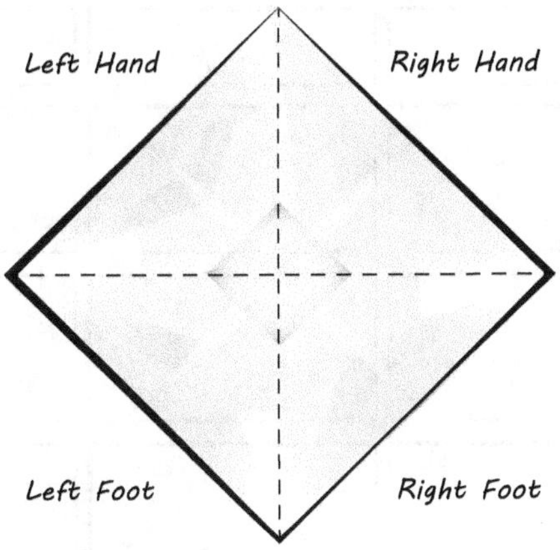
Left Hand
Right Hand
Left Foot
Right Foot

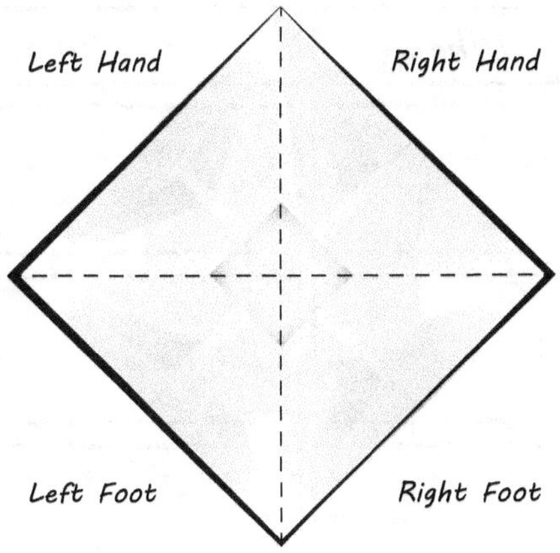
Left Hand
Right Hand
Left Foot
Right Foot

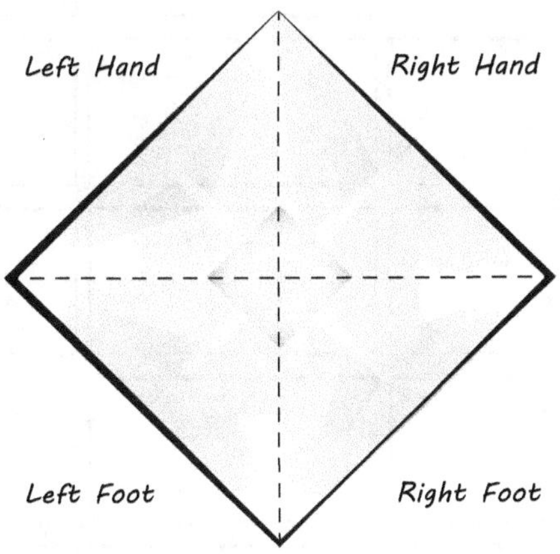
Left Hand
Right Hand
Left Foot
Right Foot

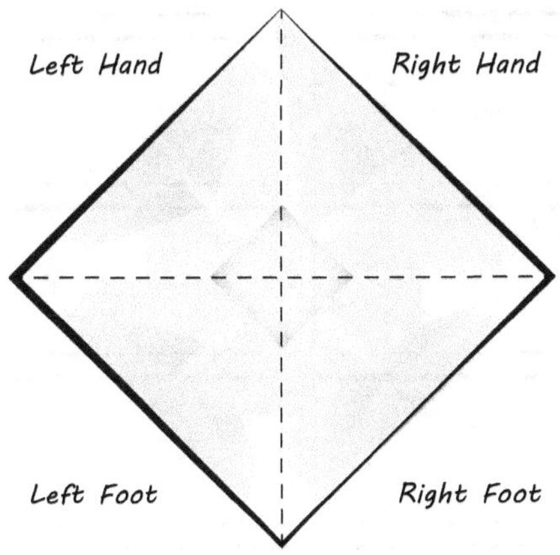
Left Hand
Right Hand
Left Foot
Right Foot

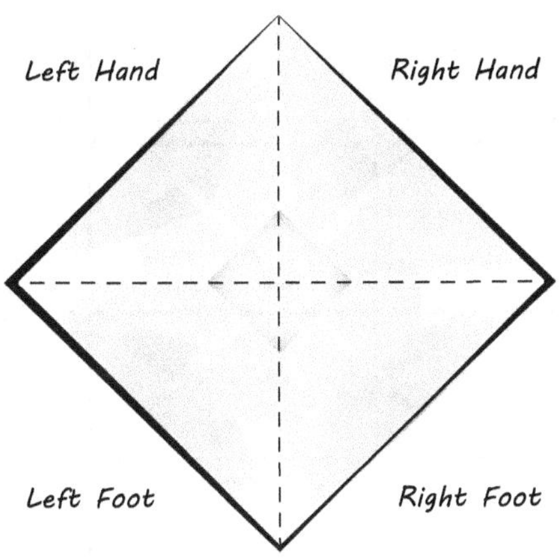
Left Hand
Right Hand
Left Foot
Right Foot

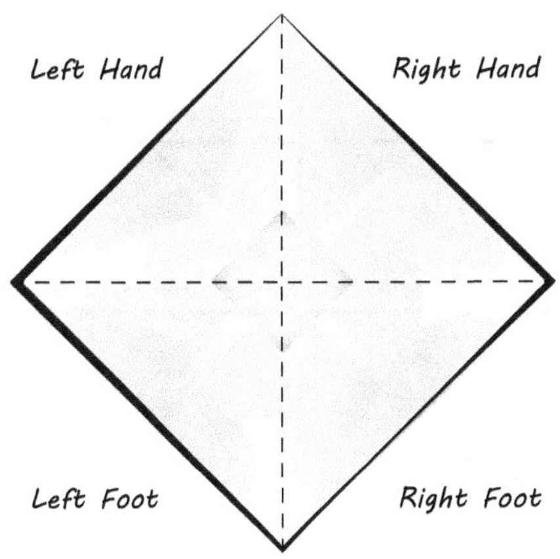
Left Hand
Right Hand
Left Foot
Right Foot

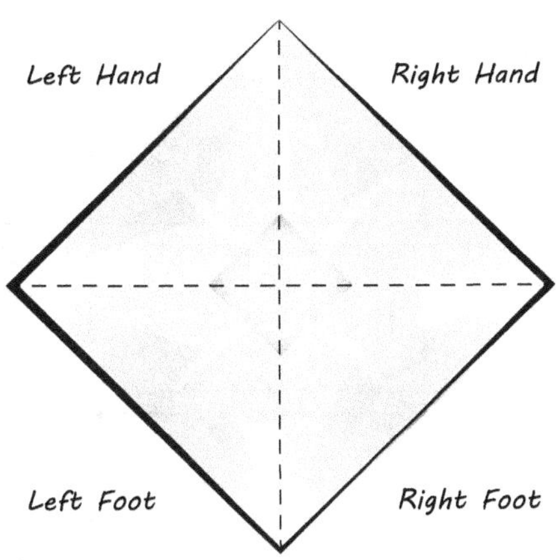
Left Hand
Right Hand
Left Foot
Right Foot

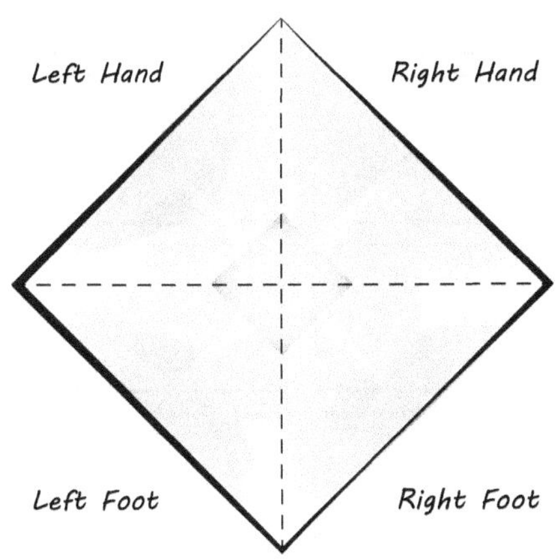
Left Hand
Right Hand
Left Foot
Right Foot

Drum groove creation

When	What	Tempo	Notes/Reflections

Chapter 15 The power of feel

Overview

At this point you should be well on your way to creating so many new and varied grooves. Well done, it's a great journey to be on and one that never really stops. But we should not underestimate the *'Power of Feel'*. This chapter aims to discuss concepts that will help establish and improve the feel within your grooves.

Things we will cover in this chapter…

- Hi-hat feel variations
- Sound substitution
- Layering for dynamics
- The mixing board
- Behind, on or in front of the beat
- Maintaining time through drum fills

Hi-hat feel variations

I wanted to add a section dedicated to hi-hat variations. So far, we have explored some of our essential hi-hat options. However, there are many other variations that can offer our grooves an alternative and varied feel and impact our music. In these examples I have added the backbeat so you can see how it would work alongside the snare drum. You could vary this, of course, and then add in your own bass drum patterns.

This is based around added or subtracted notes as well as different accents placements and patterns. Experiment on these using your created and amalgamated grooves.

1970s R&B/Soul 16ths with accents on 'ands'

This example is based on our 16th note hi-hat pattern, but with an accent on the 'and' of each beat. This will give you an alternative feel to this style of grooves. This one can be effective at slower tempos.

DOI:10.4324/9781003585169-15

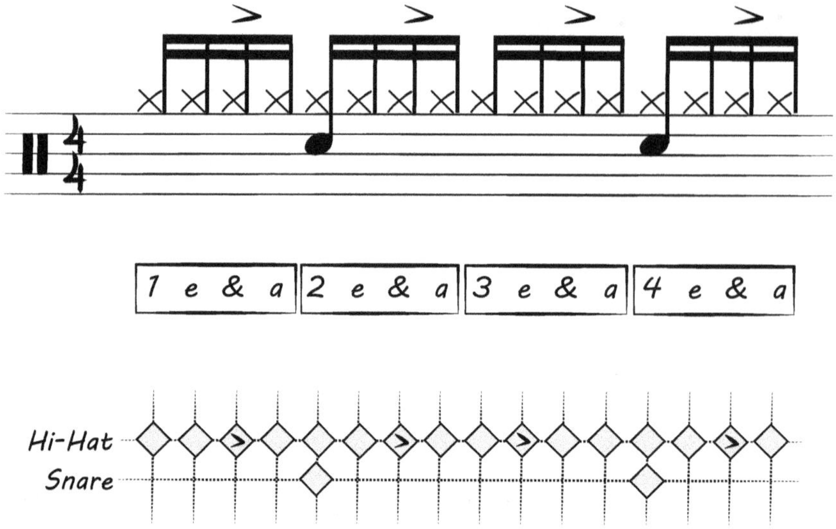

1980s drum machine hi-hats with accents

This is inspired by the hi-hats in Michael Jackson's *Wanna be Startin' Something*.

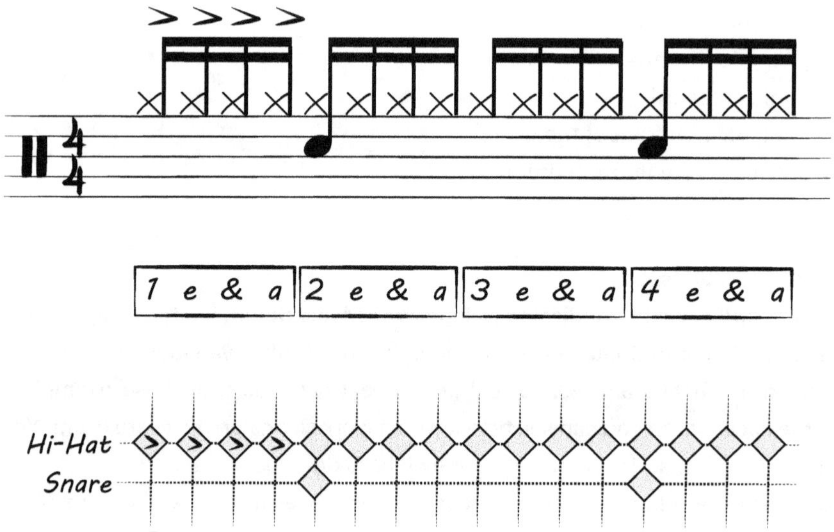

Early 1990s Gangsta Rap/Hip-Hop style

Quite a simple one here, and often created on a drum machine back in the early 1990s. One of the cool things about this, however, is that it means we are not (always) playing the hi-hats at the same time as the backbeat. This allows the snare drum to really 'pop' in the groove.

Example one

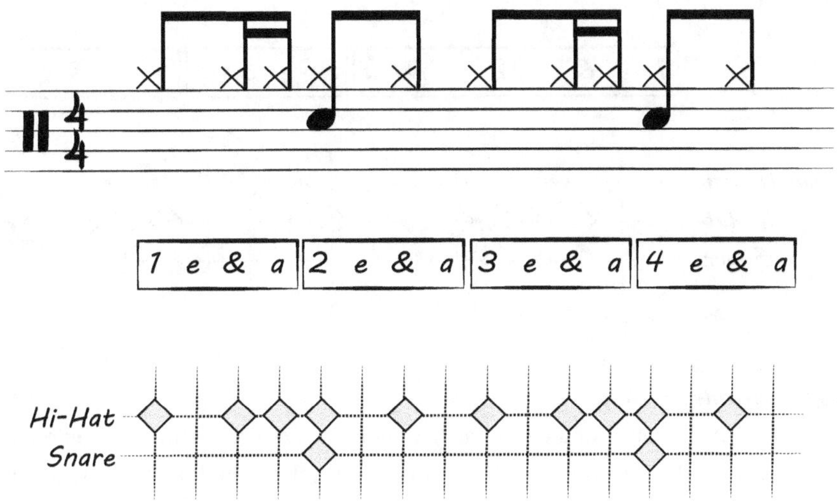

Example two

A fun little two-bar pattern inspired by *Hypnotize* by Notorious B.I.G.

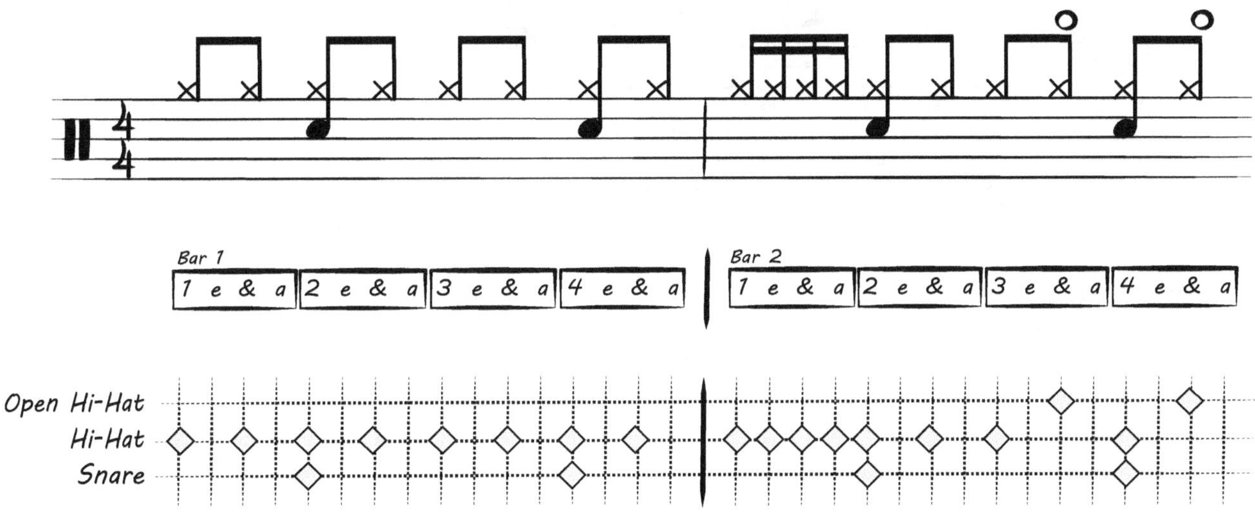

Mid-1990s house example

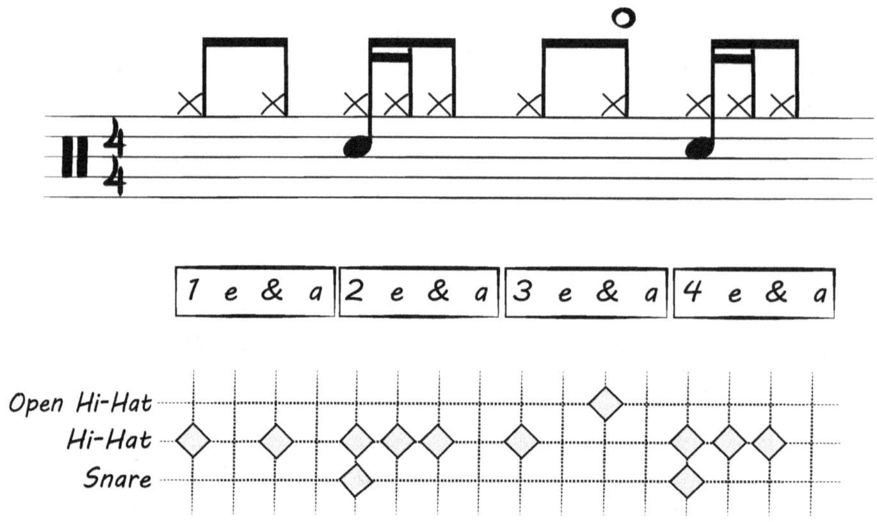

Go-Go beat hands example

Here we see a presentation of a Go-Go-style beat (hands only). In a more traditional sense, much like some of the Afro-Cuban or Brazilian grooves, there are foot patterns that often fit to a groove. For now, though, we are just taking the Go-Go style hand pattern as inspiration to be able to use over the top of other beats.

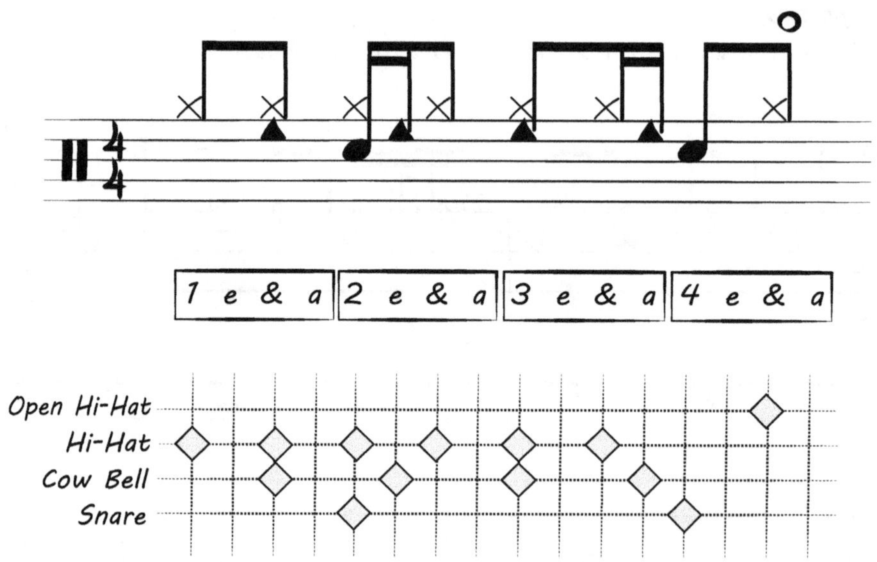

Sound substitution

At the end of Starting Essentials, I briefly discussed the concept of sound substitution. Please revisit this if you need to refamiliarise yourself with it. What we discussed was taking our timekeeper (i.e. hi-hat) and moving it onto different sound sources to create different feels and effects. Experiment with some of the different hi-hat options and patterns with this concept of sound substitution and see what you can come up with.

Here are some you can try...

- Bell of ride cymbal
- Floor tom
- Shell of the drum
- Rim of the drum
- Jingle stick
- Shaker or maracas
- Mini timbales
- Tambourine (either handheld or hit with a stick)

The influence of technology

Of course, technology allow us to layer multiple sound sources. Here are some I like to use in terms of technology:

- Hand claps on the snare beats
- Tambourine hit on the snare beats
- Layering these two in a bar so there is one sound on the first backbeat and a second sound on the sound (very Michael Jacksonesque)
- A slight delay effect on the drums (this can work well on the hi-hats or snare)

Layering and sound substitution in a track

So, this is for everyone but can be especially useful for beatmakers/producers who want to add different layers of intensity and dynamics to a song. Layering the drums with additional sounds can be a great way to do this. Try these ones out and hear the effects in action:

- Adding a tambourine on quarters, 8ths or 16ths in a chorus.
- Doubling up the backbeat with either a bass drum or floor tom to create a unison
- Changing the backbeat to a cross-stick in the verse

One last one to try. When going from the chorus back to a verse, try not to crash into the verse. Generally, the idea is that you are taking the dynamics down; adding a loud and explosive crash can have the opposite effect.

The mixing board

Imagine the Mix window in your DAW or faders on an analogue mixing desk. In a live situation or in the recording studio, it is common that each of your drums would have a microphone on them. How do we want to mix the volumes of our individual drums and cymbals to add to our feel? A good place to start with this is by lowering the volume of the hi-hats and allowing the snare and kick to stand out by comparison.

This can also be further explored by thinking of each element of the groove on a scale of 1–10 in terms of volume. Within a groove trying playing/programming each part at a different volume to find what works and what doesn't.

Behind, on or in front?

This refers to where is the groove sits in comparison to a metronome and the general pulse of the music.

Behind – this can give the music a laid-back feel or make it feel like it is 'deep in the pocket'. This can be created by placing the backbeat slightly late and just after the hi-hat note that goes with it, creating a flam between the two.

On the beat – hitting exactly on the metronome. A really great way to get people tapping their feet to your music.

In front – ahead of the beat, giving a sense of urgency and drive. This is the opposite to behind the beat in the sense that you would now place the snare just ahead of the beat. If it is playing with a hi-hat note, you would create a flam but this time one with snare (ever so slightly) before the hi-hat.

Keeping a sense of time during fills

Often it can be easy to lose a sense of feel, timing and consistency when executing drum fills. They can sound like a separate entity to the groove and almost like time has been suspended. One way around this is to keep a constant sense of pulse when a drum fill is being played. This could be through keeping quarter notes through the bar on the hi-hat pedal.

Keeping the backbeat

This can be used in addition to the hi-hat pedal on quarter notes during a fill. Whilst you don't have to actually play a backbeat, giving an accent to notes that land on two and four can help maintain a sense of pulse and time in the music and contribute to the overall feel of a drum track.

Feel in relation to other instruments

Whilst all of the above is important, how this all fits and ties in with other instruments is crucial.

Bass drum – Lock this in with the bass guitar. Do the bass drum notes complement the bass guitar, or do they clash with them rhythmically? This is what will help ground the rhythm section and create a foundation for a track. Try to keep the bass drum melody consistent and not overly complex (unless this is called for by the style of music).

Snare drum – Keep a backbeat that is easy to pick out. Keep a separation in volume between the ghost notes and snare to allow the backbeat to cut through other instruments in the track, and for the ghost notes to add additional texture. Be wary not to overuse ghost notes (perhaps if there is a particularly rhythmic guitar part that it could clash with). You don't want to be competing for space in the track.

Hi-hats – There is a fine line between playing that fits the track – and playing that is too complex. Avoid making your hi-hat part too complex or too varied. Maintain a consistent solid groove.

Groove creation: Challenge fifteen

There is a common theme emerging in terms of our groove creation challenges in these later chapters: apply the chapter's knowledge to your own grooves. It is exactly the same in the case of the 'Power of Feel' chapter. Take and experiment with the concepts discussed in this chapter and apply it to your own grooves. Don't forget to keep track of your progress.

When	What	Tempo	Notes/Reflections

Chapter 16 General tips and things to try

General tips and things to try if things aren't working for you with creating your own drum grooves

- Simplify – sometimes less is more.
- Ostinato one (being hi-hats 8th and snare two and four) is always a good place to start.
- Record your playing/creations and listen back.
- For creating in different styles – to get an authentic approach, go and listen to these different genres of music.
- Try creating to something. This could be a bass loop, rhythmic guitar part or anything really that can give you inspiration to create your own grooves.

Tips for creating grooves for a song

Up until this point we have primarily been discussing creating singular drum grooves in isolation, but a large thing left unsaid is constructing grooves for a song. Here are some things that we may think about to help give a song the thing that we think its missing.

Texture

- Perhaps switching between hi-hats on the verses and go to the ride cymbal on the choruses.
- Sometimes between verses I will lose an element of the drums to give an altered feel. This could be through leaving out the back beat.

Dynamics

- This is not just loud or quiet, but a variety of intensities demonstrated throughout the song.
- A way to do this could be through closed tight hi-hats in the verses and then splashier looser hi-hats for the choruses.
- Altering between a cross-stick and a snare drum as a backbeat in different sections.
- Orchestration of timekeeper. This is briefly mentioned in the part about exploring sound substitution but altering the surface of the timekeeper can be a cool way to add variety and flavour to your songs.

 DOI:10.4324/9781003585169-16

- A nice one can be playing the bridge on the tom-toms to have a greater impact coming back to the chorus afterwards.
- Never underestimate the power of silence on the drums. Taking the drums out of a section completely, or putting a stop or pause, can be an impactful way to add to your song.

Creating for the song

- Listen to what the other instruments are doing and how the drums can complement them. This could be through emphasising certain key hits or moments. Generally, it is useful to let the melody breathe.
- More syncopation in the verses and less in the choruses. This is not a hard and fast rule, but it can be useful in creating an anthemic sound chorus if that is what the music is going for.

When	What	Tempo	Notes/Reflections

Chapter 17 Conclusion

Well done for reaching the end of *Drum Groove Creation*. If you have followed all of the chapters, examples, steps and created your own grooves then that is a massive achievement and something that should not easily be forgotten about or pushed aside. As mentioned in the introduction, the ultimate goal behind this book is to help you learn and create your own drum grooves. But, as I am sure you can imagine, this is not the end of the journey, it is more like the beginning of it. Creating drum grooves never stops and it will keep growing as you discover new music, grooves and get inspired to make drums for different musical scenarios.

A personal note from me

In in all honesty this whole book and journey has been incredibly personal for me, and I can hand on heart say that I have loved the opportunity to write, create and publish this book to help other people out there who want to be creative and create new drum grooves.

To everyone, thank you for your love and support and for joining me on this journey through *Drum Groove Creation*. I hope to meet you one day, hear the grooves you have created and talk all things music.

Thank you and see you next time.
Guy Monk

DOI: 10.4324/9781003585169-17

Glossary

Ahead of the beat	when a groove is slightly ahead of the metronome
Amalgamation	a combination of different things
Beats per minute (BPM)	the number of pulses which happen inside a minute
Behind the beat	when a groove sits just behind the metronome; sometimes to create a laid-back feel
Downbeats	beats falling on the quarter notes (on the pulse)
Feel	a combination of timing, texture and dynamics to help lift a groove
Groove	a repeated set of notes played on the drums
Iteration	a variation of a given example
On the beat	when a groove sits exactly on the metronome
Orchestrate/orchestration	moving a musical idea or phrase onto different parts of the drum set
Ostinato	a repeated pattern
Straight	where the notes are spaced evenly within a beat
Subdivision	a value of note(s)
Swing/swung	the notes are not spaced precisely evenly within each beat; the music has a slight 'bouncy lilt' to it
Syncopation	rhythms falling, off and around the pulse
Tempo	the pulse and rate of speed of a piece of music
Time signature	the two numbers that indicate how many beats in a bar and what rate of speed to count
Upbeats	the beats that fall in between the pulse

Index

Pages in *italics* refer to figures.

Index

Index

For Product Safety Concerns and Information please contact our EU representative GPSR@taylorandfrancis.com
Taylor & Francis Verlag GmbH, Kaufingerstraße 24, 80331 München, Germany

www.ingramcontent.com/pod-product-compliance
Lightning Source LLC
LaVergne TN
LVHW082306300926
849324LV00039B/1028

* 9 7 8 1 0 3 2 9 5 4 8 5 1 *